Inadvertent Climate Modification

The MIT Press
Cambridge, Massachusetts,
and London, England

Inadvertent Climate Modification

Report of the Study of Man's Impact on Climate (SMIC)

Sponsored by
the Massachusetts Institute of Technology

Hosted by
the Royal Swedish Academy of Sciences and
the Royal Swedish Academy of Engineering Sciences

This book was designed by The MIT Press Design Department.
It was set in Linotype Baskerville
by The Colonial Press Inc.,
printed on Finch Textbook Letterpress
by The Colonial Press Inc.
and bound by The Colonial Press Inc.
in the United States of America.

ISBN 0 262 19101 6 (hardcover)
ISBN 0 262 69033 0 (paperback)

Library of Congress catalog card number: 79-170861

SAMUDRA MÈKHALÈ DEVI PARVATHA STHANA MANDALÈ PĀDA SPARSAM KSHMASWA MÈ

Oh, Mother earth, ocean-girdled and mountain-breasted, pardon me for trampling on you.

Sanskrit Prayer

Contents

Preface

The need for the 1971 Study of Man's Impact on Climate (SMIC) was perceived by several of us who had been deeply involved in the planning and conduct of the 1970 Study of Critical Environmental Problems (SCEP). SCEP, which had been sponsored by M.I.T. and had been participated in by more than 70 U.S. scientists, examined global environmental problems in some detail and published the results in October 1970. Following that intensive and extensive exercise, we concluded that an examination of the critical questions associated with the climatic effects of man's activities should be conducted by scientists from many nations at the earliest possible date.

The implications of inadvertent climate modification, both in terms of direct impact on man and the biosphere and of the hard choices that societies might face to prevent such impacts, are profound. Should preventive or remedial action be necessary, it will almost certainly require effective cooperation among the nations of the world. SMIC was developed to assist in this process by providing an international scientific consensus on what we know and do not know and how to fill the gaps. It is hoped that this consensus will provide an important input into planning for the 1972 United Nations Conference on the Human Environment and for numerous other national and international activities.

The committee that planned SMIC during the fall of 1970 and the spring of 1971 was chaired by Professor Carroll L. Wilson, M.I.T., and included Dr. William W. Kellogg, National Center for Atmospheric Research; Dean Thomas Malone, University of Connecticut; Professor William H. Matthews, M.I.T.; and Dr. G. D. Robinson, Center for the Environment and Man, Inc.

The following scientific objectives were adopted for SMIC: to review SCEP findings critically; to point to global environmental problems that were slighted or overlooked; to obtain a more complete assessment of present knowledge; to amplify certain points, especially with respect to establishing priorities among problems and determining how to proceed in implementing recommendations for action and further extensive research; and to point to questions requiring international policy decisions.

Members of the Planning Committee had numerous discussions with persons in several governmental and nongovernmental international organizations and in national agencies in several countries. As a result of these consultations, participants were selected and invited and background materials for the Study were prepared.

Thirty scientists from fourteen countries participated in SMIC for the three-week period of June 28 to July 16, 1970. The participants, predominately atmospheric scientists, represented a broad spectrum of specialities including climatology, meteorology, fluid dynamics, physics, chemistry, mathematical modeling, oceanography, and hydrology. The Study was conducted at the conference center Wijk near Stockholm, Sweden, and was hosted jointly by the Royal Swedish Academy of Sciences and the Royal Swedish Academy of Engineering Sciences.

The scientific and technical judgments of the participants on the present status of understanding of man's impact on climate are presented in this Report. It is the result of three weeks of intensive work and discussion by five SMIC Work Groups, as well as discussion and careful consideration by all participants.

This Report represents the consensus of all of the SMIC participants. Each participant concurs with the substantive presentation, but everyone did not have an opportunity to review the final wording of all chapters. Participants in the Study acted as individuals, not as representatives of the agencies or organization with which they were affiliated.

Chapter 2 of Part I of this Report presents a summary of the major conclusions and recommendations of the Study. This chapter was adopted as presented here by all of the participants and has been distributed as an official Study document prior to publication of this book. The remainder of the book appears in essentially the same form as adopted before the conclusion of SMIC. Some editing was done during the week following SMIC by Drs. William H. Matthews, William W. Kellogg, G. D. Robinson, Philip D. Thompson, and Stephen H. Schneider, but great care was taken not to alter the meanings or emphases of important points.

Because the major objective of SMIC was to raise the level of informed public and scientific discussion and action on global and regional climate problems, this Report was published as quickly as possible. Rapid publication has undoubtedly resulted in less editorial smoothness than might have been possible if individual chapters had been substantially reworked following the Study.

Parts I and II should provide an overview of the topic of inadvertent climate modification for the lay reader. Parts III and IV are, in general, rather detailed assessments of present knowledge about specific problem areas. These parts should be particularly useful to atmospheric scientists and those in related areas for they provide the perspective and detail necessary for a rather fundamental understanding of a wide range of topics.

During the planning of SMIC, warm encouragement and support were offered by several international groups, and we hope that even before this book is published they will have profited by the SMIC deliberations. Specifically, draft manuscripts were forwarded immediately to the Secretariat of the 1972 U.N. Conference on the Human Environment, to the World Meteorological Organization for an Intergovernmental Work Group on Monitoring meeting in August, to the Scientific Committee on Problems of the Environment (SCOPE) of the International Council of Scientific Unions (ICSU) for a major meeting in early September, and to the Joint Organizing Committee of the Global Atmospheric Research Program (JOC/GARP) for an October meeting. These organizations have kindly agreed to review and comment on the SMIC Report for the Secretary-General of the U.N. Conference, Mr. Maurice F. Strong, whose encouragement and interest have been of such great importance to this effort. Finally, it should be noted that the timing of the rapid publication of this Report has been dictated by the meeting of the Preparatory Committee for the U.N. Conference in late September.

Financial support for the Study was provided by grants from the National Science Foundation through the newly created program of Research Applied to National Needs (RANN), by the Ford Foundation, by the American Conservation Association, and by the Sloan Foundation.

A report of this scope and depth could not have been developed without the assistance and dedication of numerous organizations, groups, and individuals. On behalf of the SMIC participants, we wish to thank all those who contributed to this effort.

Carroll L. Wilson, SMIC Director
William H. Matthews, Associate Director

July 1971

Director
Carroll L. Wilson
MASSACHUSETTS INSTITUTE
OF TECHNOLOGY

Associate Director
William H. Matthews
MASSACHUSETTS INSTITUTE
OF TECHNOLOGY

Joint Secretaries
William W. Kellogg
NATIONAL CENTER FOR
ATMOSPHERIC RESEARCH

G. D. Robinson
CENTER FOR THE ENVIRONMENT
AND MAN, INC.

Participants

Pál Ambrózy
CHIEF OF DEPARTMENT
METEOROLOGICAL SERVICE
OF HUNGARY
BUDAPEST, HUNGARY

Jean Bricard
PROFESSOR
LABORATORY OF PHYSICS
OF AEROSOLS
UNIVERSITY OF PARIS
PARIS VI, FRANCE

M. I. Budyko, Director
PROFESSOR
MAIN GEOPHYSICAL OBSERVATORY
LENINGRAD, U.S.S.R.

Hermann Flohn
PROFESSOR OF METEOROLOGY
UNIVERSITY OF BONN
BONN, FEDERAL REPUBLIC
OF GERMANY

Hans-Walter Georgii*
PROFESSOR
DEPARTMENT OF METEOROLOGY
AND GEOPHYSICS
UNIVERSITY OF FRANKFURT
FRANKFURT AM MAIN, FEDERAL
REPUBLIC OF GERMANY

Edward D. Goldberg
PROFESSOR OF CHEMISTRY
SCRIPPS INSTITUTION
OF OCEANOGRAPHY
LA JOLLA, CALIFORNIA, U.S.A.

Hans Hinzpeter
PROFESSOR
INSTITUT FÜR METEOROLOGIE DER
JOHANNES GUTENBERG
UNIVERSITÄT
MAINZ, FEDERAL REPUBLIC
OF GERMANY

Joachim H. Joseph
PROFESSOR
DEPARTMENT OF ENVIRONMENTAL
SCIENCES
TEL-AVIV UNIVERSITY
RAMAT-AVIV, ISRAEL

Christian Junge*
PROFESSOR
MAX-PLANCK INSTITUT FÜR
CHEMIE
MAINZ, FEDERAL REPUBLIC
OF GERMANY

William W. Kellogg, Director
LABORATORY OF ATMOSPHERIC
SCIENCE
NATIONAL CENTER FOR
ATMOSPHERIC RESEARCH
BOULDER, COLORADO, U.S.A.

Tatsumi Kitaoka, Director
METEOROLOGICAL RESEARCH
INSTITUTE
JAPAN METEOROLOGICAL AGENCY
TOKYO, JAPAN

Julius London
PROFESSOR
DEPARTMENT OF ASTRO-
GEOPHYSICS
UNIVERSITY OF COLORADO
BOULDER, COLORADO, U.S.A.

* SMIC Work Group Leader

Robert A. McCormick, Director
DIVISION OF METEOROLOGY
NATIONAL OCEANIC AND ATMOSPHERIC ADMINISTRATION
ENVIRONMENTAL PROTECTION AGENCY
RALEIGH, NORTH CAROLINA, U.S.A.

Lester Machta, Director*
AIR RESOURCES LABORATORIES
NATIONAL OCEANIC AND ATMOSPHERIC ADMINISTRATION
SILVER SPRING, MARYLAND, U.S.A.

Syukuro Manabe
GEOPHYSICAL FLUID DYNAMICS LABORATORY
NATIONAL OCEANIC AND ATMOSPHERIC ADMINISTRATION
PRINCETON, NEW JERSEY, U.S.A.

William H. Matthews
ASSISTANT PROFESSOR
DEPARTMENT OF CIVIL ENGINEERING
MASSACHUSETTS INSTITUTE OF TECHNOLOGY
CAMBRIDGE, MASSACHUSETTS, U.S.A.

Jacques Van Mieghem
PROFESSOR OF METEOROLOGY
UNIVERSITY OF BRUSSELS
BRUSSELS, BELGIUM

Fritz Möller
PROFESSOR
METEOROLOGISCHES INSTITUT
UNIVERSITY OF MUNICH
FEDERAL REPUBLIC OF GERMANY

R. E. Munn
DEPARTMENT OF THE ENVIRONMENT
TORONTO, CANADA

Jehuda Neumann
PROFESSOR AND HEAD
DEPARTMENT OF METEOROLOGY
HEBREW UNIVERSITY
JERUSALEM, ISRAEL

Marcel Nicolet, Director
PROFESSOR
INSTITUT D'AÉRONOMIE SPATIALE DE BELGIQUE
BRUSSELS, BELGIUM

P. R. Pisharoty
PROFESSOR
PHYSICAL RESEARCH LABORATORY
AHMEDABAD, INDIA

G. D. Robinson
CENTER FOR THE ENVIRONMENT AND MAN, INC.
HARTFORD, CONNECTICUT, U.S.A.

R. W. Stewart, Director*
PACIFIC REGION, MARINE SCIENCES BRANCH
DEPARTMENT OF THE ENVIRONMENT
VICTORIA, BRITISH COLUMBIA, CANADA

Karoly Szesztay
SECTION HEAD
RESEARCH INSTITUTE FOR WATER RESOURCES DEVELOPMENT
BUDAPEST, HUNGARY

Philip D. Thompson, Associate Director*
NATIONAL CENTER FOR ATMOSPHERIC RESEARCH
BOULDER, COLORADO, U.S.A.

Sean Twomey
DIVISION OF METEOROLOGICAL PHYSICS
CSIRO
SYDNEY, AUSTRALIA

Ottavio Vittori
PROFESSOR
ISTITUTO DI FISICA DELL'ATMOSFERA
LABORATORIO CONSIGLIO NAZIONALE DELLE RICERCHE
BOLOGNA, ITALY

* SMIC Work Group Leader

Carroll L. Wilson
PROFESSOR
SLOAN SCHOOL OF MANAGEMENT
MASSACHUSETTS INSTITUTE OF TECHNOLOGY
CAMBRIDGE, MASSACHUSETTS, U.S.A.

Giichi Yamamoto
PROFESSOR
GEOPHYSICAL INSTITUTE
TOHOKU UNIVERSITY
SENDAI, JAPAN

Consultants

Willi Dansgaard
PROFESSOR
H. C. OERSTED INSTITUTE
UNIVERSITY OF COPENHAGEN
COPENHAGEN, DENMARK

Erik Eriksson
PROFESSOR OF HYDROLOGY
UNIVERSITY OF UPPSALA
UPPSALA, SWEDEN

Research Staff

Stephen H. Schneider
GODDARD INSTITUTE FOR SPACE STUDIES
NEW YORK, NEW YORK, U.S.A.

Joseph C. Perkowski
MASSACHUSETTS INSTITUTE OF TECHNOLOGY
CAMBRIDGE, MASSACHUSETTS, U.S.A.

Support Staff

Constance D. Boyd
THE M.I.T. PRESS
CAMBRIDGE, MASSACHUSETTS, U.S.A.

Allan Carlsson
ROYAL SWEDISH ACADEMY OF ENGINEERING SCIENCES
STOCKHOLM, SWEDEN

Ada B. Demb
MASSACHUSETTS INSTITUTE OF TECHNOLOGY
CAMBRIDGE, MASSACHUSETTS, U.S.A.

Francis H. McGrory
MASSACHUSETTS INSTITUTE OF TECHNOLOGY
CAMBRIDGE, MASSACHUSETTS, U.S.A.

Inadvertent Climate Modification

Part I Overview and Summary

1 Introduction

1.1
The Focus of SMIC

This Study has provided an authoritative assessment of the present state of scientific understanding of the possible impacts of man's activities on the regional and global climate. Based on this assessment, specific recommendations have been developed for programs that will provide the knowledge necessary for more definitive answers in these complex areas. The following major areas were considered by SMIC:

—Previous climate changes
—Man's activities influencing climate
—Theory and models of climate change
—Climatic effects of man-made surface change
—Modification of the troposphere
—Modification of the stratosphere

For each of these topics, the following general questions were addressed:

—What can we now say authoritatively on the subject?
—What are the gaps in knowledge that limit our confidence in the assessments we can now make?
—What must be done to improve the data and our understanding of their significance so that better assessments may be made in the future?
—What programs of focused research, monitoring, and/or action are needed?
—What are the characteristics of action needed to implement the recommendations of the Study?

1.2
The Present Status of Knowledge

During the past two decades, there has been significant and encouraging progress by the scientific community in developing the theory, models, and measurement techniques that will be necessary for determining man's impact on climate. We are, however, disturbed that there are major and serious gaps in our understanding of the complex systems that determine climate, and that data in many critical areas are incomplete, inconsistent, and even contradictory.

It has been possible for us to identify some parts of the climate system, such as the arctic sea ice, that seem particularly sensitive to the impact of man's activities. We have urged that high priority be given to efforts that would significantly improve our understanding of these potential problems. In other areas, however, we can conclude only that we do not yet know enough to make positive assertions about man's potential role in climate change.

It is clear to us that without additional research and monitoring programs the scientific community will not be able to provide the firm answers which society may need if large-scale, and possibly irreversible, inadvertent modification of the climate is to be avoided.

We feel that the information that would be obtained through the implementation of the major SMIC recommendations would fulfil these needs. Provided with such information, a future Study of Man's Impact on Climate could be far more definitive about the effects and impacts of man than we have been able to be.

1.3
The Audiences for the SMIC Report

The general, and indeed ultimate, audience of the SMIC Report is comprised of the peoples and governments of the world who collectively bear the responsibility for assuring that man does not inadvertently destroy his environment in the process of meeting his many and varied needs. It is hoped that this Report, particularly through its summary in Chapter 2 and through Part II, will provide a clear and understandable "progress report" to concerned and intelligent laymen about the status of present knowledge of man's impact on climate. More specifically, this Report is addressed to the several audiences whose understanding, acceptance, and implementing action are needed if further progress is to be made in this critical area.

A primary audience is the scientific community. We hope that these analyses, conclusions, and recommendations are persuasive to interested scientists all over the world. Effective implementation of these recommendations will require an international consensus among concerned scientists if the research is to be conducted that will ultimately answer critical questions more definitely than we can today. We also hope that the SMIC Report will stimulate talented young scientists to develop their career interests through study of the fascinating and complex problems of climate change and man's influence on it.

Our second audience includes those national officials whose decisions govern the allocation of resources to support scientific programs and the directors of those institutes and laboratories where programs must be initiated, expanded, or modified to implement these recommendations. We hope that such persons will consider that the implementation of these recommendations carries a sufficiently high priority to justify allocation of the resources needed.

Our third audience is comprised of those international organizations whose encouragement was vital in carrying out this Study (note discussion in the preface), and whose response to our recommendations and whose integration of our recommendations into their programs will be so important in achieving early action at all levels. These include the World Meteorological Organization (WMO), the Scientific Committee on Problems of the Environment (SCOPE) of the International Council of Scientific Unions (ICSU), and the Joint Organizing Committee of the Global Atmospheric Research Program (JOC/GARP). A focal point for this exercise has been the 1972 United Nations Conference on the Human Environment because it will provide a unique forum for consideration and action on these and related recommendations.

Our fourth audience includes those officials of national and international organizations that provide technical and financial assistance to developing countries. Some of the SMIC conclusions and recommendations are of special concern to these countries, and they will require assistance of various kinds in order to act on these recommendations. Their participation in many of the research, measurement, monitoring, and policy evaluation activities is essential for the success of global networks. Furthermore, such close involvement is essential for the development of independent judgments in the formulation of national policies and programs.

A fifth audience is comprised of those organizations and

groups whose deliberations focus on the institutions needed to implement these recommendations. We hope that those in this audience, and particularly the delegates to the 1972 U.N. Conference on the Human Environment, will give special attention to those SMIC recommendations that call for the coordination of national programs, the agreement on standards of measurement, the selection of sites for measurement and monitoring, and the identification of appropriate forums for evaluation of trends that may indicate the need for concerted action by many countries.

1.4 Some Financial Implications of the SMIC Recommendations

The primary focus of the scientists at SMIC was on substantive scientific questions. There was neither the time nor the expertise to determine the costs of the various research programs that are needed if man is to understand his impact on climate. Many of the SMIC recommendations, however, closely parallel those of the SCEP Report, and at the latter Study a work group was able to devote some thought to estimation of costs. Thus the SCEP Report provides a rough approximation of the financial implications of the SMIC recommendations.

A major consideration in estimating the cost of an international program of research and measurement is to establish the rules of accounting. For example, continuing measurements which require skilled or semiskilled operation of delicate equipment are sometimes required at remote locations where shelter and services must be provided. Some existing meteorological or telecommunication stations could accommodate some of the new work without inconvenience or great additional cost. The question arises, however, of what proportion of the cost of the shelter and services should be charged to the new venture. The same question arises when ship and aircraft facilities are required. At SCEP, it was assumed that where this multiple use of facilities was possible only extra costs would be charged to the new project. The same assumption governs the very rough cost estimates to follow.

Considerable research will be required to implement the SMIC recommendations for the development of the theory of climate and for computer simulation of man-induced changes of climatic factors. We believe that much of this research can be undertaken through continued support and in some cases expansion of the efforts of already-existing groups, especially if there is constant attention to maintaining close and continuing communication among these groups. There are, for example, at least five very capable research groups working on the problem of modeling the general circulation of the atmosphere. These multiple efforts provide the independence necessary at this stage of development for encouraging new formulations and fresh approaches to these problems. The annual budget for a highly trained scientific and technical team with the most advanced computers is on the order of $2.5 million, including the amortized capital outlay for the computer.

SMIC has noted the need for ten measurement stations in very remote areas, and it must be assumed that buildings will be required for some. The capital cost of each such station is not likely to be less than $0.5 million and the annual maintenance charge for each will be on the order of $0.25 million.

Before embarking upon an extensive oceanographic research and monitoring program related to climate studies, SMIC has recommended that preliminary ocean surveys be undertaken to provide the necessary data for the design of a major program. To keep two oceanographic ships at sea and provide a modest staff to study their findings would cost about $4 million each year. This figure assumes no new shore establishments.

The SMIC recommendations for a global network of about 100 stations are essentially the same as those of SCEP. These stations, providing systematic samples of air and also precipitation chemistry and solar radiation measurements, would be supported by central laboratory and standardizing facilities. Here extensive use of existing facilities is possible because meteorological and other scientific services already operate the kind of facility required and indeed have in some cases already undertaken these additional activities. It seems reasonable to estimate an additional annual cost of $1 million, with a similar capital cost for equipment. It is more difficult to cost the additional aircraft facilities required by this program. The performance of aircraft available on commercial lines is sufficient for much of the work, and some has already been accomplished on routine passenger-carrying flights. Another $1 million per year seems to be a reasonable estimate.

The program of stratospheric research and observation proposed by SMIC calls for the use of high-performance aircraft and for many launchings of large balloons. Costing the flying time of military or prototype supersonic aircraft was not possible at SMIC. The total figure for a five-year program might be on the order of $30 million, but this is only a very rough estimate. This is a very small sum compared to that already spent on the development of the aircraft that fly, or will fly, in the stratosphere.

Finally, the use of artificial earth satellites as platforms for monitoring and research equipment is mentioned in several recommendations. If simple additions on existing meteorological or other satellites are assumed, instrumental costs of the order of $0.5 million per year might be expected.

The total cost, roughly estimated, of the climatic research and monitoring program outlined in SMIC would probably be about $17.5 million per year, with a capital sum of $7 million primarily for establishment of remote stations. This figure compares well with that produced more carefully, but for a somewhat different program, by SCEP.

1.5
Some Institutional Implications of the SMIC Recommendations

The costs involved in implementing the SMIC recommendations and the time delays in taking action depend largely upon which institutions—national and/or international—ultimately take the recommended actions of measurement, monitoring, research, and evaluation and interpretation for policy decisions. SMIC has attempted to be definitive and authoritative in assessing the status of our present knowledge and in recommending needed programs. However, SMIC has refrained from making any specific recommendations about the institutions that should undertake these activities, because we feel that such decisions will and must be based on considerations in addition to those in which this Study is expert. We can, however, offer a few general observations about the institutional implications of the SMIC recommendations.

Implementing many of these recommendations requires action through national and international programs and institutions. If the proposed research, measurement, and monitoring programs could be undertaken only by new or as yet nonexistent organizations, or by the enlargement of existing organizations, the costs would be large and the delays would be long before actions could begin. Fortunately, most of these recommendations can be implemented by additions to or modifications of existing programs, and through existing laboratories and institutes where there are already a large fraction of the scientific and technical staff needed. However, in cases such as the establishment of new monitoring stations, new programs, institutions, and staffing will be needed.

We also see an important role for international bodies in promoting coordination of national programs, in the establishment of agreed standards for measurement and monitoring, and in the evaluation of results and encouragement of widespread dissemination of information. We attach great importance to the identification of the appropriate international forums in which there can be a continuing assessment of those activities of man which may have a serious impact globally or in large geographic regions. Through these forums agreements should be sought for common national policies and programs that will avoid or reduce the impacts which may jeopardize the globe or large regions.

2 Major Conclusions and Recommendations

2.1 Previous Climate Changes

2.1.1 Discussion

The climate of a locality is defined by long-term statistics of the variables of state of the atmosphere at that locality. Conventionally the climate is represented by the mean value and the variance of a long series (usually 10 to 30 years) of observation of those variables. There are local and regional climates and a planetary or global climate. The regions chosen are generally the natural geographical regions of the earth's surface. Short time variations of meteorological parameters entail "weather changes," while their long-term variations entail "climate changes." The climate may be considered as the sum of a set of weather types, each of these types having a characteristic frequency distribution over the averaging period.

Any general discussion of man's impact on climate should begin with a review of the observed fluctuations of climate not only during the period of instrumental observations (since about 1680) and the period of abundant historical, botanical, and geological data since the maximum of the last ice age (about 20,000 years before present) but also since the beginning of higher organized life on our globe about 550 million years ago. In relation to the earth's history, our recent climatic fluctuations, covering less than 300 years, are only small-scale noise, are seemingly random, and are certainly not well understood.

Perhaps the most significant lesson to be learned from the long history of our planet is that during more than 90 percent of this 550 million year period the earth was free of polar ice. In a sense, we live in an ice age, and this is an anomaly for our planet. During the late Paleozoic period, 250 to 300 million years ago, an ice age occurred that lasted 30 to 50 million years. Following this there was a long period without any polar ice, and then there was a gradual cooling that resulted in the beginning of the present antarctic glaciers during the Pliocene period, about 5 million years ago. The glaciers in the Northern Hemisphere apparently began much later, first in the Sierra Nevada of California and Iceland about 2.5 million years ago and soon thereafter in Greenland.

In this period of glaciation in which we find ourselves now there have been several periods which we define as the Pleistocene, during which large ice sheets formed over the North American and European continents. The retreat of these Pleistocene ice sheets took place only 8000 to 16,000 years ago, so recently that its effects on the rocks and the distribution of soil are still clearly visible. A remarkable feature of this retreat is the relative abruptness with which it took place. Evidence from the Greenland ice sheet, which has persisted and preserved the record in the layers of snow that accumulated each year, suggests a rapid increase of the mean temperature by several degrees C—just how many degrees we do not know—during a period of about 1000 years at the end of the last ice age. For the last 10,000 years the temperature, as indicated by the Greenland snow, has been relatively constant compared with the earlier dramatic change. There have, however, been many significant smaller fluctuations, such as the cold period in the seventeenth century that has been referred to as "the little ice age."

Since accurate thermometers became available, we have been able to make more precise records of our climate, and we know that the mean temperature of the Northern Hemisphere (for which the better data exist) increased by about 0.6°C from 1880 to 1940, and since 1940 has fallen by more than 0.3°C. These mean temperature changes are misleading, however, since it is very significant that in the last 30 years the changes in warming and cooling in arctic regions were three times larger than the mid-latitude changes, and in the tropics there was little change, possibly even a slight change in the opposite direction. While it is conceivable that man may have had a small part in the most recent climate changes we have just described, it is clear that natural causes must be sought. In fact, as has been frequently pointed out, it will be difficult to identify any man-made effect because, first, with our present state of knowledge, we do not know how to relate cause and effect in such a complex system and, second, man-made effects will be obscured by the natural changes that we know must be occurring.

2.1.2
Recommendation

We recommend that scientists from many disciplines be encouraged to undertake systematic studies of past climates, particularly of climates in epochs when the Arctic Ocean was free of ice (see Section 3.3.7).

2.2
Man's Activities Influencing Climate

One of the most clearly evident influences of man on his environment is his pollution of the atmosphere. In cities the pollution is sometimes acute, but even in the most remote places in the world it is still possible to detect traces of man-made contaminants in the air. The resulting rise in the particulate load of the atmosphere can be attributed to both industrial activity and the burning of waste crops and vegetation that is a practice in many tropical areas. Particles scatter and absorb solar radiation and also have an effect on the outgoing infrared radiation from the surface, so these man-made products will influence the heat balance over wide areas.

Some gaseous constituents of the atmosphere also absorb solar and infrared radiation, and carbon dioxide (CO_2), water vapor (H_2O), and ozone (O_3) are in this category. It is well known that the CO_2 content of the global atmosphere has been rising due to the burning of fossil fuels—coal, petroleum, and natural gas—and it is expected that it will go up about another 20 percent by 2000 A.D.

Until quite recently it has been assumed that man could not compete directly with nature in the release of heat on a large scale, but now we must take a further look at this matter as we realize the implications of a doubling of the present world population of 3.6 billion by the year 2000, coupled with an expectation of more energy to be used per capita. The production of energy of all sorts is rising at a rate of from 5 to 6 percent per year for the world (5.5 percent per year is equivalent to a factor of 5 in 30 years, or by the year 2000 A.D.). There may eventually be industrialized areas of 10^3 to 10^5 km^2 where the additional input of energy by man will be equivalent to the net radiation from the sun; and on a continental scale the present insignificant contribution may rise to 1 percent of the continental net radiation average after about 40 years.

We have spoken of some of the effects of industrial activity, but for thousands of years before the Industrial Revolution agricultural and animal grazing practices have had a profound influence on large regions of the world, and it seems very likely that these have already resulted in changes of climate of those regions. Grazing by goats and other domestic animals has reduced parts of Africa and Southwest Asia to semideserts; dense forests of the mountainous areas from Turkey to Afghanistan and of the Mediterranean, Europe, and the eastern United States have been cut down to make arable or grazing land; and the savannah grasslands of the tropics are nearly all man-made. The net result is that some 20 percent of the total area of the continents has been drastically changed, with a consequent change in the heat and water budget. In arid or semiarid areas the growing demands for water for irrigation have reduced the reserves of groundwater, and the process of irrigation increases the water vapor in the air.

Another influence of man is his manipulation of the surface waters by building dams, creating lakes, draining swamps, and diverting rivers. Artificial lakes, for example, will alter the heat balance of the area because water has a lower albedo; it has a much greater heat capacity, and it adds water vapor to the air.

Perhaps the diversion of rivers from one region to another has even greater potential implications, since the water so diverted will convert substantial dry desert or semidesert dry areas to irrigated farmland, and three-fourths to nine-tenths of irrigation water is evaporated in the air.

Control of river discharge into ocean areas subject to winter freezing could greatly influence the rate of freezing or melting. Such activities coupled with intentional dusting of sea ice to hasten melting could have serious regional and even global repercussions.

Modification of precipitation by seeding clouds with freezing nuclei has been sufficiently attractive to encourage the practice throughout the world. Any change in precipitation pattern influences the heat budget of the atmosphere; therefore, widespread seeding operations, including efforts to change the course of hurricanes, may modify the normal patterns of rainfall and snowfall enough to have an influence on the heat budget of a part of the atmosphere.

Finally we come to one of the most rapidly escalating activities of man, his various modes of transportation. Automobiles contribute approximately half of some air pollutants observed in U.S. cities, and the same is true in most other industrialized countries. Since the automobile exhaust forms "smog" particles, this is a major contribution to the particle increase that we have already discussed. It is also interesting to note that roads in the United States cover nearly 1 percent of the entire area. This is not as big a change of land use as that due to agriculture, which has been discussed earlier, but it is not entirely trivial.

Aircraft also produce exhaust products, and it is currently estimated that commercial aviation will double its jet fuel consumption every five or six years in the next decade or so. These exhaust products from jet airliners are deposited high in the troposphere or the lower stratosphere. There are indications that jet traffic has already caused a small increase in cirrus cloudiness in heavily traveled areas, and this will have a small effect on the heat balance of the atmosphere. The supersonic transports fly in a region where the average residence time of their exhaust products is one or two years, so there is a chance for the concentration to build up.

In speaking about the future of man and his climate we are faced with two important classes of consideration: meteorological (or geophysical) and social. Whereas we do have a rudimentary theory of climate and a good deal of knowledge about how things behave in the physical world, when we turn to a forecast of man's behavior we must resort to simple extrapolation of present trends with only slight shadings to take account of the more obvious interactions that we can foresee. There are no laws that we know of, or mathematical models, that will allow us to predict the future course of human affairs better than such extrapolations. Therefore, we can in the end only forecast what *could* happen *if* mankind proceeds to act in a certain way, more or less as he is acting now.

2.3
Present Theory and Models of Climate Change
2.3.1
Discussion

Climate is determined by a balance among numerous interacting physical processes in the oceans and atmosphere and at the land surface. Locally and globally, climate is subject to change on all scales of time, but the physical processes themselves remain the same and are amenable to study by statistical, physical, and mathematical techniques. If we are to assess the possibility and nature of a man-made climatic change, we must understand how the physical processes produce the present climate and also how past changes of climate, clearly not man-made, have occurred.

Several physical-mathematical techniques—they have come to be called "models"—are being developed to attack the problem. We have distinguished four types: (1) "global-average" models, in which horizontal motion of the atmosphere is neglected; (2) parameterized semiempirical models that consider the whole atmosphere and surface but simulate some of the effects of atmosphere and oceanic motions with the aid of empirically adjusted constants; (3) statistical-dynamical models in which physical laws are applied to statistics of the atmospheric variables; and (4) explicit numerical models in which motions and interactions are treated in detail by integrating mathematical equations expressing the time rate of change of the variables, though in practice the detailed treatment must be abandoned at some minimum scale of motion and replaced by empirical parameterization or statistical methods.

The global-average models serve the purpose of identifying and investigating in a preliminary way problem areas of concern to climate theory; for example, they have been used to make first estimates of the effects of changing CO_2 content. The semiempirical models allow consideration of problems of climate. There may be some uncertainty about their indications because we do not understand in detail the parameterized processes or the consequences of parameterization; nevertheless, their predictions must not be ignored, particularly if, as is often the case, the indications of independently developed models are similar. Statistical-dynamical models are in an early stage of development, but it is hoped that they will at some later stage obviate some of the enormous and costly computational load associated with explicit numerical models. Significant progress has been made in developing these detailed models, enough to give appreciation of the magnitude of the task and confidence that it might succeed.

We now know enough of the theory of climate and the construction of climatic models to recognize the possibility of man-made climatic change and to have some confidence in our ability ultimately to compute its magnitude. Recent results obtained using empirical models have in our view increased the urgency of the study of climate theory in its own right. Some of these results, for example, concern the delicate balance of the processes that maintain the arctic sea ice. The empirical models suggest that a small change in mean air temperature or in solar radiation reaching the surface could result in considerable expansion or contraction of the ice pack—a climatic change of great significance to human life. This in our opinion reinforces the urgency of studies of climate by all available methods in order to understand its natural and possible man-made changes.

2.3.2
Recommendations

Data must be gathered judiciously for the understanding of climate change and for the verification of the accuracy of atmosphere-ocean climate models. We therefore recommend:

1. Monitoring the temporal and geographical distribution of the earth-atmosphere albedo and outgoing flux over the entire globe, with an accuracy of at least 1 percent (see Section 5.2.4).
2. Monitoring with high resolution the global distribution (horizontal and vertical) of cloudiness, and the extent of polar ice and snow cover with less resolution (see Sections 5.2.4, 6.5.6, and 6.8.6).
3. Measuring the distribution, optical properties, and trends of atmospheric particles and clouds over the globe (see Sections 5.2.4 and 6.8.6 and Chapter 8).

4. Determining the absolute value of the solar constant to better than ±0.5 percent and the spectral distribution of solar radiation from 1800 Å to 40,000 Å to a few percent (±1 percent in the visible part of the spectrum) (see Section 5.2.4).

5. The surface temperature of the ocean should be monitored if the information is obtainable by remote sensing from satellites (see Sections 5.3.4 and 6.8.6).

We would like to be able to recommend a monitoring program for the temperature distribution and currents in the upper ocean. However, we recognize that there is at present no economical and effective way to perform such monitoring. Instead, therefore, we recommend:

6. Combined theoretical and observational studies to determine the best way to obtain the oceanic data required to verify joint ocean-atmosphere models (see Sections 5.3.4 and 6.8.6).

Joint atmosphere-ocean models should be refined and extended to evaluate the magnitude of climate change resulting from man's activities and to differentiate it from natural climate change. We therefore recommend:

7. New and improved joint atmosphere-ocean models that incorporate the effect of the following (see Section 6.8.6):

a. Cloudiness (where preferably the cloudiness is generated by the model itself).

b. Sea ice and snow cover (where the extent and thickness of the ice is predicted by the model).

c. Sea-air interface exchange of heat, moisture, and momentum, and the turbulent exchanges in both boundary layers.

d. Particles in the atmosphere.

Man's impact should be studied by testing the effect on climatic changes of carbon dioxide and other trace gases, aerosols, heat released, and surface changes. We therefore recommend:

8. The expanded use of simplified parameterized climate models in order to gain insight into some of the basic factors of climate and climate change. These models can be used to test the impact of the individual factors on climate and require relatively little computing time (see Sections 6.4, 6.6, and 6.7).

9. Long-term integration of a realistic atmosphere-ocean cli-

mate model to test for long-period climatic fluctuation inherent in the natural system (see Section 6.2.2).

2.4
Effects of Man-made Surface Changes
2.4.1
Discussion

We have considered those aspects of the interaction between the atmosphere and the surface of the planet which play a part in controlling the climate in the context of both natural climatic changes and of the possibility of intentional and inadvertent interferences by man. Some of the activities of man which cause important surface changes were noted earlier in this summary.

The most sensitive surface state appears to us to be the ice and snow, particularly arctic sea ice, because of the large change of albedo accompanying a change in its area and the relative ease of modification. Inadvertent change in the next four decades will be mainly of regional importance. However, over the only somewhat larger time scale of the next century we recognize a real possibility that a global temperature increase produced by man's injection of heat and CO_2 into his environment may lead to a dramatic reduction or even elimination of arctic sea ice. Further, there is a possibility that deliberate measures to induce arctic sea ice melting might prove successful—and might prove difficult to reverse should they have undesirable side effects.

The example of the arctic sea ice is an interesting illustration of the sensitivity of a complex and perhaps unstable system that man might significantly alter over the next few decades by making relatively small modifications in the earth's present heat balance. Some models indicate that a few degrees C change in average temperature of the Northern Hemisphere might begin a melting of the arctic sea ice.

Some studies indicate that large areas of open arctic sea would tend to cause melting of the remaining sea ice, with the eventual result that it would disappear entirely. Once gone, it probably would not readily freeze over again. The melting of the arctic sea ice would not affect ocean levels. The changes in climate—

especially in the Northern Hemisphere—which would occur after the arctic sea ice melted are unknown, but they might be large and include changes in precipitation, seasonal temperatures, wind systems, and ocean currents. The possible effects on the Greenland ice cap are likewise unknown—whether it would increase because of an increase in precipitation or begin to melt, and how long it might take for such changes to occur, although we believe any such changes would take many centuries. A start has been made in studying, by the use of models, the sensitivity of arctic sea ice to temperature changes.

With respect to man's impact on land surfaces, we have considered possible effects of the heat release resulting from all man's activities in energy conversion—"useful" as well as "waste" energy eventually finds itself heating the environment. We see the possibility of climatically significant changes on a regional scale in the near future. On the local scale, this influence is already very large. Local and regional scale effects are amenable to study by existing climatic models.

Finally, we have considered man's manipulation of groundwater and his use of surface waters. These produce large local climatic changes, for example when an irrigated region becomes cooler. On a global scale the effects are principally upon albedo and on altering the amount of evaporation that takes place. We suggest that these effects could be studied with the use of existing models, but it seems that most of the work remains to be done. In particular, the climatic impact of deforestation of tropical jungle, which seems to result in sharply reduced evaporation and increased local heating, requires study. We note that "mining" of fossil groundwater is leading to loss of a nonrenewable resource, and note that this and inadvertent tapping of other groundwater supplies may be depleting them to the extent that they may be causing the recent rise in sea level.

2.4.2
Recommendations

1. We recommend that an international agreement be sought to prevent large-scale (directly affecting over 1 million square kilometers) experiments in persistent or long-term climate modifi-

cation until the scientific community reaches a consensus on the consequences of the modification (see Section 7.3.1).

2. We recommend that model experiments be undertaken to reveal the consequences of altering the ratio of direct heating of the surface and of evaporation, for example, in marginal semi-arid regions (see Section 7.4.7).

3. We recommend that a climatological census be undertaken, to keep a record of such sensitive indicators of state of the climate as arctic sea ice, glacier mass, and sea level, as well as the intensity of human activities likely to influence climate such as fuel consumption, area under irrigation, and urban area (see Section 7.5).

4. We recommend that the work of those attempting to study climatic variation be aided by making readily available to them seasonal regional averages of relevant data such as area-averaged precipitation records (see Section 7.1.2).

2.5 Modification of the Troposphere

2.5.1 Discussion

We have identified certain ways in which man is modifying the composition of the lower atmosphere or troposphere. His activities are increasing the number of particles in the atmosphere by adding them directly and by adding gases such as SO_2 which react in the atmosphere to form particles. By burning fossil fuels he is also increasing the CO_2 content of the atmosphere. We also suspect that man's activities may be changing the cloudiness of the atmosphere in a significant manner, for though this is not yet definitely established it is obvious that there are local man-made clouds.

Studies, particularly some using "global average" climatic models, show that particles affect the climate through two major processes. They directly affect the transfer of radiative energy in the atmosphere, and they modify the properties and the processes that lead to rain and snow. Changes of CO_2 content also modify the transfer of radiative energy in the atmosphere, and we shall discuss this later.

The extent to which present and potential man-made changes

of atmospheric composition have a significant effect on the climate is still in doubt, because of gaps in our knowledge of the magnitude of the man-made change, of the processes of emission, transformation, and removal which regulate this magnitude, and of the relevant physical and chemical properties of the added material.

There is a considerable population of "natural" particles in the atmosphere, a fact that complicates attempts to estimate man's contribution. We find wide divergences in current estimates of the total particle population, of the proportions existing in various geographical locations, and in the contributions from the various sources. We are not able to resolve these divergences.

We find there is insufficient knowledge of the optical and chemical properties of the particles to allow full use even of the diagnostic "global-average" models of climate, and there are insufficient observations of the radiative properties of polluted atmosphere to allow verification of the results of theory and computation. However, calculations based on reasonable estimates of refractive index and particle content indicate that low-level layers will cause a net cooling. Our recommendations for research and measurement are generally similar to, but more comprehensive than, those in the SCEP Report.

Theoretical studies of radiative transfer in the atmosphere show that cirrus clouds can have a very important influence on atmospheric processes. We find that little is known about the formation, persistence, and optical properties of cirrus clouds, and that we are unable to estimate the extent to which they might be affected by man's modification of the atmosphere.

We find the situation in regard to the effects of CO_2 on climate a little clearer than for particles. Evidence that has appeared since the SCEP study appears to confirm that the rate at which CO_2 is being accumulated in the atmosphere fluctuates, though the average increase continues to be about 0.2 percent per year. Thus, the natural processes that regulate the proportion of man-made CO_2 which remains in the atmosphere must vary in effectiveness from year-to-year. We agree with the conclusion of SCEP that these should be studied, but we are less ready than was SCEP to recommend specific measurements. In our opinion, more detailed preliminary examination of the problem of ocean-atmosphere partition of CO_2 is required before a continuing measurement program for the CO_2 content of ocean waters can be recommended.

We have reconsidered the question of the oxygen content of the atmosphere in light of the material published since the SCEP study but see no reason to modify the SCEP conclusion that there is no danger of significantly depleting atmospheric oxygen as a result of burning fossil fuels.

2.5.2 Recommendations

We recommend the following activities in research and monitoring as having the highest priority for assessing the effects on climate by particles and gases that man introduces into the troposphere:

1. *Direct influence of particles on the atmosphere:*

a. Better figures on global production rates of particles and reactive trace gases (see Section 8.3.4).

b. Studies of the transformation of trace gases to particles (see Section 8.3.4).

c. Studies of the radiation field of short- and longwave radiation in clean and polluted atmospheres (see Section 8.7.7).

d. Studies of the refractive index of atmospheric particles (see Section 8.7.7).

e. Monitoring global distribution and time trends of particles by measuring turbidity, total particle concentration, ice nuclei, and cloud nuclei (see Section 8.6.4).

2. *Influence of particles on clouds:*

a. Studies of the change of albedo of clouds due to pollution (see Section 8.7.7).

b. Comprehensive field studies on the change of cloud cover, type of cloud, and precipitation caused by pollution (see Section 8.7.7).

c. Develop objective methods for monitoring cloud with high precision and reliability by satellites (see Sections 5.2.4 and 8.7.7).

3. *CO_2 and water vapor:*

a. Monitoring CO_2 at about 10 stations (see Section 8.8.7).

b. Monitoring H_2O in the upper troposphere and lower stratosphere (see Section 8.8.7).

c. Study formation process of cirrus clouds (see Section 8.9.4).

d. Develop suitable methods for monitoring trends in cirrus cloud amount (see Section 8.9.4).

2.6
Modification of the Stratosphere
2.6.1
Discussion

The region of the atmosphere from about 15 to 50 kilometers above the earth is the stratosphere; below this is the troposphere. For the first time in history man now can put material directly into the stratosphere in the form of combustion products such as water vapor, nitric oxide, other gases, and particles from the exhaust of high-flying aircraft. The SCEP analysis of the climatic consequences of these products using an assumed SST fleet size, operations, and engine emissions concluded the following:

1. There may be an increase in stratospheric cloudiness in polar regions in the winter, but neither the amount of the change nor the climate consequences are known.
2. Radiative effects due to locally injected amounts of carbon dioxide, carbon monoxide, water vapor, and nitric oxide would be small compared with those due to other man-made sources of these gases.
3. The sulfur dioxide, hydrocarbons, and nitric oxide would produce particles that may play a role in the atmospheric heat balance as they do when introduced by volcanoes, but the consequences on surface temperature are unknown.
4. Ozone changes will be small compared with the present day-to-day and geographical variability of total ozone.*

No new information has been developed to alter appreciably the SCEP judgment on these issues.

Since SCEP, the role of decreasing ozone and consequent increasing ultraviolet radiation that may damage life, including man, has been emphasized.

Ozone in the stratosphere is important and partly controls stratospheric temperature and shields life on earth from harmful ultraviolet radiation. The ozone concentration in this region is

* We believe that this fact is not relevant in considerations of man's impact on climate.

maintained by balance between several competing chemical reactions and atmospheric transport processes. There is uncertainty in some of these reactions and in the complete understanding of the transport processes involved. Nor do we know much about the transport of products from the troposphere that might affect ozone concentrations.

Ozone concentration measurements at many places in different parts of the world have shown upward trends during the past decade. These are believed to be natural fluctuations, but we cannot explain why they occur or be certain that transport of man-produced materials from the troposphere has not had some effects.

We do not believe that present data and knowledge permit drawing any firm conclusions as to the effects of supersonic transport combustion products on ozone concentrations in the stratosphere. We consider that answers to these questions should be produced before large-scale aircraft operation in the stratosphere becomes commonplace, and we believe that solutions might be produced by concentrated research.

2.6.2
Recommendation

The main needs for understanding possible climatic modification due to changes in stratospheric constituents either of natural origin (such as from volcanoes) or from man-made sources (such as from high-flying airplanes) are a field survey and routine monitoring of upper troposphere trace gases and particle concentrations. Aircraft and balloons are existing sampling platforms, but in some cases, new measurement techniques must be developed. Seasonal, geographical, and altitude climatology and long-term trends must be part of the program. For some problems, particularly for stratospheric chemistry, laboratory measurements of uncertain rate coefficients of chemical reactions are needed.

An equally important and difficult problem is the stratospheric transport of ozone and other trace gases. Both chemical reactions and transport processes must be incorporated in the realistic models to resolve uncertainties in the prediction of effects of constituent changes (see Section 9.8).

Part II

Climate and Man in Perspective

3 Fundamentals of Climate Change

3.1 The Purpose of SMIC

The atmosphere-ocean environment of the Planet Earth has slowly evolved over a period of more than 4 billion years. Even a cursory comparison of Earth with its neighboring planets, Venus and Mars, reveals the fact that our remarkable long-term balance between oceans, dry land, and atmosphere is not necessarily the only state that our planet could maintain. What other courses could the evolution of Earth have taken under slightly different circumstances (Rasool and DeBergh, 1970)? We cannot say, of course, but by the same token we have no assurance that our balance is stable. In fact, the history of the planet, recorded in the rocks and sediments of the ocean floor, suggests that the balance may have nearly been upset many times in aeons past.

One product of this planetary evolution has been mankind, and "modern man" may be said to have walked the Earth for about 50,000 years. A few thousand years ago he began to achieve a degree of mastery of his local environment, and within the past few centuries he has been changing the face of the land, the composition of the atmosphere, and the balance of living things.

It is not the industrialized parts of our civilization alone that have brought about these changes, for the impact of man's agricultural practices, by the sheer magnitude of his ever-increasing numbers and the demands he sometimes places on fragile biological systems, have degraded large areas of the land, created near deserts, and eliminated many species of plants and animals. As the world population continues to increase from the present 3.6 billion people to perhaps twice that many by the year 2000 A.D., and his ability to convert natural resources to useful products and services also increases, these demands on the environment will escalate even faster than the population.

How far can this trend go? How much can we push against the balance of nature before it is seriously upset? There are many dimensions to this balance, and in this Study of Man's Impact on the Climate (SMIC) we shall explore only what we can say now about effects on the average physical conditions in the atmosphere and about some of the closely related effects on the oceans and polar ice. These effects may cause changes in what we know as "the climate," defined rather broadly, as will be seen.

A few things are already clear. For example, we can detect worldwide or hemispheric changes in carbon dioxide and airborne particles attributable to man's activities; and we can see that man is adding enough heat over substantial areas to change the heat available to the atmosphere by several percent. All of these influences are increasing with time; for example, in the industrialized regions of the world the heat being generated will probably double every 10 to 15 years for the next few decades at least. Will these affect the climate of the earth? Certainly they will to some degree, but the uncertainty lies in the fact that we are competing with powerful natural forces that are also causing the climate to change, as it has changed in the past. We must understand these forces well enough to be able to put man's activities into proper perspective relative to nature's. That is our aim.

We have a conviction that mankind *can* influence the climate, especially if he proceeds at the present accelerating pace. We hope that the rate of progress of our understanding can match the growing urgency of taking action before some devastating forces are set in motion—forces that we may be powerless to reverse. Fortunately, the atmosphere-ocean system seems to be sufficiently ponderous and to possess enough inertia so that we probably have time to obtain a much better understanding before serious changes occur, but we must certainly devote more effort to the task than it has received in the past. Unfortunately, the machinery through which effective international action could be taken is also ponderous—in fact, in some cases we shall first have to invent such machinery, and this may take some time too.

The exercise we have begun would be fruitless indeed if we did not believe that society would be rational when faced with a set of decisions that could govern the future habitability of our planet. Nearly two hundred years ago Emanuel Kant in the *Critique of Pure Reason* summarized what it means to behave in a rational way: ". . . the whole interest of reason, speculative as well as practical, is centered in the three following questions: (1)

What can I know? (2) What ought I to do? (3) What may I hope?" In the matter of assessing man's impact on the climate we are, unfortunately, still grappling with the first of these as we start to reach out for the second.

3.2
Definition of Climate

Since we shall be devoting ourselves to matters of climate and climate change, it is important at the outset to define what we mean. The climate of a locality is usually defined by the long-term statistics of the variables that describe "the weather" at that locality, such as temperature, rainfall and snowfall, cloudiness, winds, and so forth. Meteorologists conventionally represent the climate by the mean values and the variances of these variables, and usually insist that the record on which such statistics are based run for at least 30 years.

The climate can be referred to one locality, or we may speak of regional and planetary or global climate, in which we use weighted averages of the climate determined at many places.

A more sophisticated way of looking at the climate, and sometimes a more informative one, is to consider the weather as a succession of complex "weather types" (which can be described in a variety of ways), and then to speak of the climate as a summation of a set of weather types, each of these types having a characteristic frequency distribution over the averaging period. Defining the climate this way permits some judgments to be made about how the variables interact with each other and this in turn gives some insight into the large-scale atmospheric circulations that govern the climate of a region. We may refer to this approach as "dynamic climatology," to distinguish it from the simpler statistical approach that is usually used.

3.3
Brief Outline of Climatic History

Any general discussion of man's impact on climate should begin with a review of the observed fluctuations of climate, not only during the period of instrumental observations (since about 1680) and the relatively recent period of abundant historical, botanical, and geological data but also the longer period since the beginning of higher organized life on our globe, about 550 million years ago. Our discussion of climatic variations—their trends, oscillations, and vacillations—thus includes a considerable spectrum of time scales: decades to centuries during the instrumental period and since the beginning of the written historical record in Europe and East Asia (about 2000 years ago), 10^3 years during the postglacial period, 10^4 to 10^5 years during the Pleistocene, and 10^6 to 10^8 years during the distant geological past. In relation to the earth's history, our recent climatic fluctuations, covering less than 300 years, are only small-scale noise, are seemingly random, and are certainly not well understood.

3.3.1
Climates of the Geological Past: 550 Million to 50 Years Ago

A rational understanding of the paleoclimatic history of the earth from the geophysical viewpoint begins with the spreading of the ocean floor and its consequences: continental drift, and the gradual shift of the earth's axis of rotation and of the poles to their present positions. The speed of all these motions varied around 1 to 5 centimeters per year. The increasing amount of paleomagnetic data has verified, at least in substance, these concepts; no attempt can be made here to describe the present status of the discussion.

The very early history of the earth, from its formation 5×10^9 years ago up to about 500 Myr ago (1 Myr = 10^6 years), is not sufficiently well known to permit a discussion of paleoclimates. Several more or less widespread glaciations may have occurred, the oldest and most extended one taking place before the onset of the Cambrian, 550 Myr ago. At that time the development of higher organized life occurred rapidly, simultaneously with a rise by several orders of magnitude in the free oxygen content of the atmosphere.

During more than 90 percent of the geological past since the Cambrian polar regions must have been ice-free in both hemispheres (Schwersbach, 1961). However, during the late Paleozoic (300 to 250 Myr ago), a widespread glaciation lasting 30 to 50 Myr occurred nearly simultaneously in South America, South and Central Africa, India, Australia, and Antarctica, which at that time formed one giant continent ("Gondwana") extending over about 65×10^6 km^2 around the South Pole. Large-scale glaciations apparently are possible only when and where large continents occupy a polar position for a period of sufficient length (Fairbridge, 1961).

Reliable paleotemperature records are available only since the Mesozoic era (about 200 to 60 Myr ago). They indicate an annual surface temperature of 8° to 10°C near both poles and 25° to 30°C in the tropics. Based on a simple circulation model, the latitude of the subtropical belts could be estimated to be 50° to 60°C (Flohn, 1964), considerably poleward of their present locations, with a large Hadley circulation extending to mid-latitudes, with seasonally varying rainfall and probably a simple Ferrel-type polar vortex. This interpretation seems to be consistent with all available paleontological evidence.

3.3.2
Initiation of the Pleistocene Ice Ages

At the beginning of the Tertiary era (60 Myr ago) the North Pole was situated north of eastern Siberia (near 78° N, 152° E) from whence it shifted very gradually into its present position. At the end of the Tertiary, during the last 10 Myr, the positions of the continents and of the axis of rotation were about the same as today. During the Tertiary the principal climatic feature was the very gradual decrease of temperature in mid-latitudes as shown in Figure 3.1, in contrast to virtually no change in the tropics. This can perhaps be attributed to the gradual shift of Antarctica to its present position, which produced a very slow increase of winter snow and mountain glaciation on that continent and a cooling of the bottom water of the oceans (Flohn, 1969). Evidently the beginning of the present antarctic glaciers occurred during the Pliocene about 5 Myr ago. The very slow cooling of the Northern Hemisphere can be explained as a result of the slow mixing processes within the oceans, together with some change in the atmospheric circulation and radiation balance. It resulted in the formation of mountain glaciers in the Sierra Nevada (California) and Iceland, about 2.5 Myr ago, and soon thereafter in Greenland, all of which certainly occurred after the formation of a large continental ice sheet in the Antarctic.

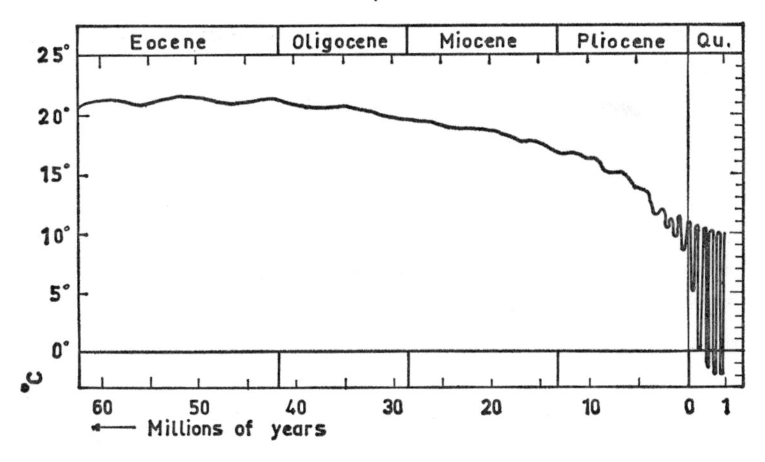

Figure 3.1 Mid-latitude temperatures during the Tertiary and Pleistocene periods relative to the present mean temperature.
Source: Flohn, 1969.

3.3.3
The Pleistocene: Glacials, Interglacials, and Pluvials

The last million years of the earth's history has been characterized by a sequence of several cold (glacial) periods with large continental ice sheets, notably in the Northern Hemisphere, alternating with warm periods (interglacials) in which the climate of the mid-latitude belt was 1° to 3°C warmer than today. During the ice ages a substantial portion of the hydrosphere was stored as huge ice domes on the continents; consequently the world ocean level lowered to about −100 meters during the last ice age 20,000 years ago and to −145 meters in one of the earlier ice ages. During the interglacials, the ocean level rose to at least 15 to 30 meters above the present level. It may have reached 70 to 80 meters at earlier interglacial periods. Evidence of still higher levels is only on a regional or local scale, affected by tectonic uplift. A vast amount of geological information has been collected by R. F. Flint (1957) and P. Woldstedt (1958–1965), but the flow of new, sometimes conflicting, evidence is still rising. The biological record has been emphasized by Frenzel (1967), and much significant evidence has been derived from deep-sea cores (Emiliani, 1955, 1966; Broecker and van Donk, 1970) and ice cores (Dansgaard et al., 1971).

The Pleistocene time scale is still a controversial question that only very recently seems to be resolved (Broecker and van Donk, 1970). The same uncertainty surrounds the number of ice ages and interglacials. At least four, more probably five, major glaciations are known to have occurred, each building up over a period of some 90,000 years and terminating in the comparatively short time of about 10,000 years. This asymmetric primary cycle is modulated by secondary oscillations with a time scale of 20,000 to 30,000 years during the glacial growth phase and of about 1000 years during the retreats.

The distribution of ice during the last ice age in the Northern Hemisphere was to some extent parallel to more recent snow cover in severe winters. Three large ice sheets have been inferred: above North America, centered in the low-lying lands around Hudson Bay; above northern Europe and Northwest Siberia, centered around the northern part of the Baltic; and above Greenland, which has apparently remained nearly unchanged since the last glacial maximum. The total area during the last ice age was 32×10^6 km^2, with an average thickness of 1200 meters. The vertical extension of the ice domes probably reached 4000 meters above sea level, similar to the antarctic ice sheet, which has also changed little since the last maximum. It is now established beyond any doubt that the last glacial maximum occurred simultaneously 18,000 to 20,000 years ago in both hemispheres (Woldstedt, 1965). The largest ice sheets were concentrated on each side of the Atlantic where the southern margin reached latitude 38° to 39° N near St. Louis and 48° to 50° N on the lower Don. The extension of the ice in the Pacific sector was comparatively small, especially in northeastern Siberia (Figure 3.2). Some coastal regions of the Arctic Ocean were also hardly affected, and the lowlands of northern Alaska and parts of the Canadian Archipelago and of northern Siberia have remained unglaciated, apparently because of a lack of precipitation.

The initial nucleus for glaciation is assumed to have occurred mainly in the mountains of northern Labrador, southern Greenland, and central Norway, all of them south of the Arctic Circle. Other areas of strong mountain glaciation, such as southern Alaska

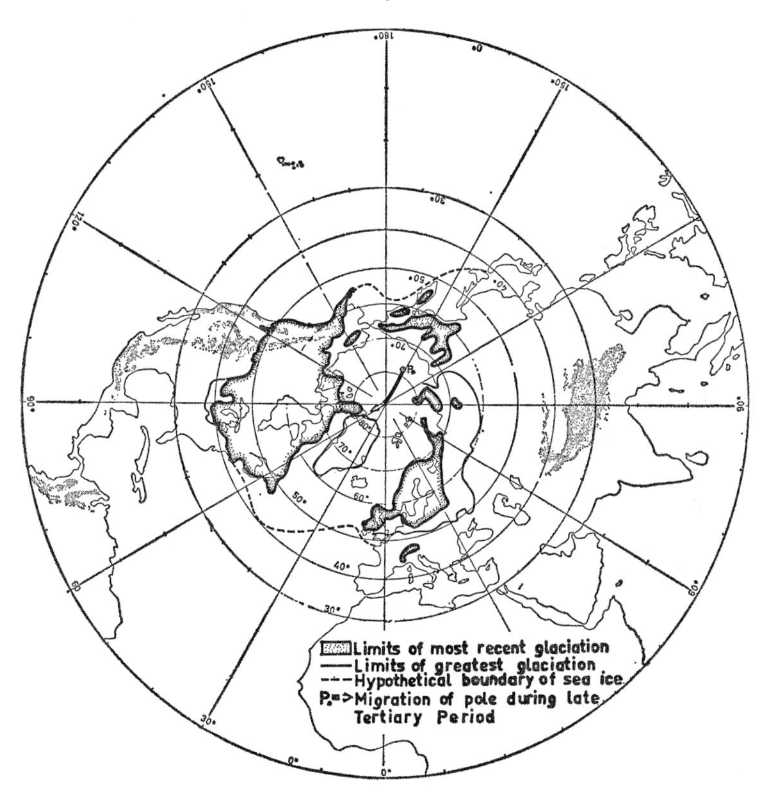

Figure 3.2 Extent of the Pleistocene ice sheets and sea ice in the Northern Hemisphere and the migration of the North (Geographic) Pole during the preceding period.
Source: Flohn, 1969.

or the Pamir-Hindukush-Karakoram area in central Asia initiated only a limited extension of glaciation. In most mountain areas the snow line was lowered by 1000 to 1400 meters, even in the tropics by 700 to 900 meters, indicating a quite general drop of temperature of 5° to 6°C. In the vicinity of the ice domes glacial katabatic winds (due to air flowing downslope and being cooled by contact with a cold surface as it goes) lowered the temperature by up to 12°C and more, compared with recent climates. The same is also true along the coasts from Ireland to France and northern Spain, in remarkable contrast to the Pacific Coast of North America, where at the same latitudes (and also in southern Chile) the temperature drop was only 5° to 6°C, as it was in many parts of the tropics.

These differences between the Pacific and Atlantic sectors can be related (Flohn, 1969) to the early closing of the Bering Strait after a sea-level drop of only 40 meters, which prevented the flow of comparatively warm Pacific water into the Arctic Ocean, while the exchange between the Atlantic and Arctic Oceans remained essentially unmodified.

In subtropical and tropical latitudes the mountain snow line, tree line, and other indirect climatic parameters dropped 600 to 1000 meters, and in some local areas in an older glaciation by even as much as 2000 meters (for example in Japan and in subtropical China). In many arid areas signs of a more humid climate ("Pluvials") are frequent, especially in northern Africa and the Middle East. The Pluvials at both flanks of the Sahara may be of quite different origin. Many apparent Pluvials were due only to the decrease of evaporation with lower temperatures, without marked change of precipitation; evidence of an actual increase of rainfall is rare.

Convincing evidence has been given for a marked arid period near the maximum phase of each glaciation (Frenzel, 1967). This may be merely a consequence of the low surface temperatures of the tropical oceans (Emiliani, 1966), when, with constant wind speed and relative humidity, the evaporation dropped by 30 or 35 percent and the net radiation was mainly used for the warming of the polar water masses of the North Atlantic (Flohn, 1969). It was during this arid phase when the widespread deposits of loess (finely divided surface material carried by the wind and deposited over a long period) were formed; they are now one of the main sources of mineral dust in the atmosphere.

During the warm periods of the interglacials the climate was essentially similar to our present climate. The antarctic ice sheet has obviously survived all interglacials; a similar statement for the Greenland ice, however, cannot be given with certainty. Europe, Siberia, and North America were somewhat warmer, at least during summer (3° to 5°C) and more humid than today, as was the Mediterranean (Frenzel, 1967).

3.3.4
Hypotheses concerning the Origin and Termination of Ice Ages

Only a few remarks on current hypotheses of the origin of ice ages seem to be justified here. The distribution of the ice cover does not seem to be consistent with the hypothesis that an ice-free Arctic Ocean was the initial condition of an ice age (Ewing and Donn, 1956, 1958); more recently evidence has been found indicating that the central Arctic Ocean probably has not been ice-free during the last 150,000 years (Ku and Broecker, 1967).

The circulation of the atmosphere during the maximum of the last ice age has been frequently studied (Willett, 1949; Lamb, 1966; Flohn, 1969). It closely resembles recent cold-winter circulation, for example, those of 1946, 1947, 1962, 1963, or 1968 to 1969. This is due to the response of the tropospheric temperature pattern to the high surface albedos of snow and ice initiating a strong feedback mechanism. Once a baroclinic frontal zone is established along the southern margins of an extended snow or ice region, the snowfall within that region is increased, which in turn tends to reestablish the baroclinic cyclogenetic zone. After the formation of a sufficiently extended and persistent snow and ice cover, this type of tropospheric circulation pattern soon becomes predominant. It is difficult to suggest an atmospheric process able to destroy this powerful feedback mechanism. Hoinkes (1968) has proposed an interesting hypothesis: during the arid phase of glaciation maximum, loess sediments were deposited right on the continental ice sheets, lowering substantially the albedo of the ice sheets and initiating rapid melting. This natural air pollution effect was certainly zero at the Antarctic and only very small at Greenland.

A quite unusual hypothesis has been suggested by A. T. Wilson (1964): a partial instability of the antarctic ice sheet as a possible trigger of the recurrent glacial-interglacial cycle. Even after some modifications of this hypothesis (Flohn, 1969) we must still await the results of a search for supporting evidence before definite conclusions can be drawn about this theory.

One of the factors that could be of importance to the development of quaternary glaciations is the change of the orbital parameters of the earth. Following the well-known studies of Milankovich (1930), the influence of orbital parameter variations on glaciations has been discussed by many authors. Numerous attempts to verify Milankovich's ideas by empirical studies of glaciation during the past million years gave contradictory results. A more direct way to gain understanding of the influence of astronomical factors on glaciation is to construct numerical climatic models. The first models of this kind (Kutzbach, Bryson, and Shen, 1968; Shaw and Donn, 1968) did not include the feedback mechanism of the ice cover's albedo, which is discussed in Chapter 6 of this book. Application of these models showed that the change of orbital parameters is insufficient to account for the development of Pleistocene glaciation. A new climatic model that takes into account the influence of polar ice albedo on climate (Budyko and Vasischeva, 1971) gave more satisfactory results in explaining the relationship between glaciations and these astronomical factors.

3.3.5 Postglacial Climatic History

The history of climate during and after the recession of the large continental ice sheets in North America and Europe is characterized by several marked fluctuations with a time scale of 10^2 to 10^3 years. Some of them are of an intensity which surpasses that of the most recent fluctuations. As a general rule, these radiocarbon-dated fluctuations occurred nearly simultaneously in both the Northern and Southern Hemispheres.

This is especially true (Heusser, 1966) for the drastic variations between 12,000 and 10,000 years before the present (B.P.). The still cool and relatively wet climate of the "Older Dryas" period was suddenly replaced by a worldwide mild, even warm, period, the "Alleröd," which in North America coincided with the "Two Creeks" period. Even more dramatic was the subsequent catastrophic readvance of the ice masses about 10,800 B.P., killing living forests wholesale in a period of probably less than a century (Lamb, 1966, p. 172). This "Younger Dryas" period created once more a tundra vegetation in wide areas of Europe. The summer temperatures had risen about 6°C at the beginning of the Alleröd;

now they dropped again 6°C and also about the same amount in southern Chile (Heusser, 1966). The sharp rise is recorded in the level of inland lakes (Broecker and van Donk, 1970), in many tropical deep-sea cores (Emiliani, 1966), and in the ice cores of Greenland and Antarctica (Dansgaard et al., 1969, 1971) (as shown in Figure 3.3). Apparently at this period a sudden change in the general atmospheric circulation occurred, simultaneously with a worldwide sharp increase of ocean temperatures.

If we disregard some minor fluctuations during the recession of the large continental ice sheets, which disappeared completely about 7000 B.P. in Scandinavia and about 5000 B.P. in northern Canada, the climax of the postglacial warming was reached between 6000 and 5000 B.P. This was once more a worldwide phenomenon, where the annual temperatures were 2° to 3°C warmer, even in Alaska more than 1°C warmer, than today. This period has been defined as the "Postglacial Optimum" or "Hypsithermal." Its mild climate, together with the relative dryness in large areas of North America and the U.S.S.R., suggests a poleward displacement of the subtropical anticyclonic belt. The arctic sea ice had receded well north of its present position, but there exists no evidence for a complete disappearance of its central area north of about 80° latitude. At the same time (and after) the Sahara and the arid parts of the Near East were considerably more humid; this means that in now completely barren arid areas there was a steppe vegetation produced by occasional severe rainstorms. Some evidence also exists for a northward extension of the tropical summer rain belt (Lamb, 1966). During this time many smaller mountain glaciers completely disappeared, and the snow line was about 300 meters higher than today. The sea level gradually rose to its present level but not above it (Shepard and Curray, 1967); after this date it was mainly controlled by the mass budgets of Antarctica and Greenland.

The following millennia were characterized by a sequence of varying climatic conditions, mostly cooler, especially during the period 950 to 400 B.C. (Lamb, 1966). The mild conditions of the postglacial optimum were nearly reached once more during the

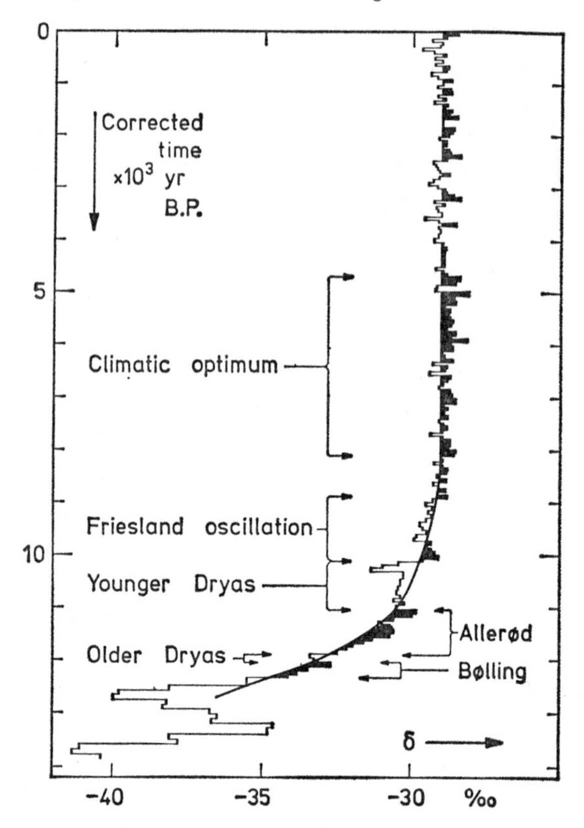

Figure 3.3 Variations of the (O_{18}/O_{16}) oxygen isotope ratio with depth below the surface (converted to age of sample) in an ice core taken from the Greenland ice sheet at Camp Century. An increase of this ratio accompanies an increase in temperature at middle and high latitudes in the Northern Hemisphere, and the values exceeding the smoothed curve (shaded in black) would therefore correspond with periods warmer than the long-term trend. The climatic periods shown refer to the European sequence.
Source: Dansgaard et al., 1971.

Table 3.1
Average Climatic Conditions over England and Wales

Dates (approx.)	Epoch	Mean Temperatures (°C) Summer (July–Aug.)	Winter (Dec.–Feb.)	Annual	Annual Rainfall (mm)	Annual Evaporation (mm)
1901–1950 A.D.	Recent	15.8	4.2	9.4	932	497
1550–1700 A.D.	Little Ice Age	15.3	3.2	8.8	867	467
1150–1300 A.D.	Little Optimum	16.3	4.2	10.2	960	517
900–450 B.C.	Subatlantic	15.1	4.7	9.3	960–979	482

Source: After Lamb, 1966.

early Middle Ages (800 to 1200 A.D.), when the ice conditions around Iceland and Greenland were much less severe than today, permitting the early Viking settlements in Iceland and Greenland and the exploration of the Canadian Archipelago and the coasts of North America with their small boats. Annual mean temperatures in southern Greenland must have been 2° to 4°C above present averages, in Europe more than 1°C (Table 3.1). This and the subsequent deterioration of climate until the seventeenth century greatly affected the cultural and economic history of Europe.

The period between 1300 and 1650 A.D. (in Norway and Alaska it extended up to 1750) brought a marked widespread change to much cooler conditions, its climax occurring between 1610 and 1640 (Lamb, 1966). The term "Little Ice Age" has been generally accepted. During this time many smaller mountain glaciers were newly formed, and all glaciers readvanced substantially, a process referred to as "neoglaciation." Yet this 400-year period could never be compared with the great ice age lasting several tens of thousands of years. It can be partly related to the period of high volcanic activity, as shown in Figure 3.4. At many mountain glaciers, nearly simultaneous advances culminated around 1640, 1740, 1820, and 1850, after which a worldwide recession of glaciers began. Temperature estimates for the several climatic periods of this time have

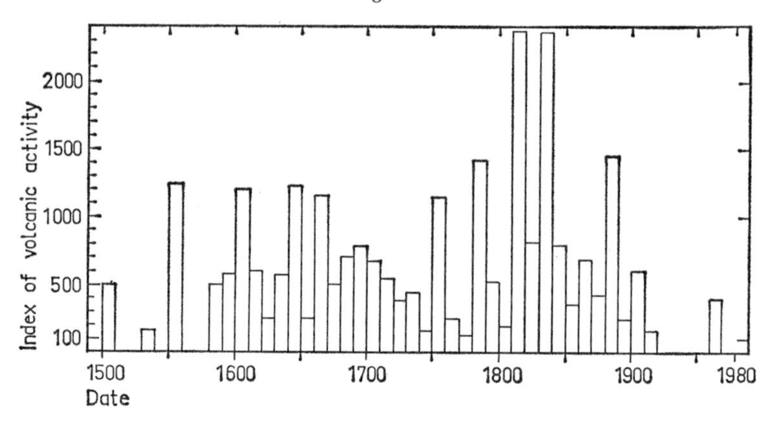

Figure 3.4 An index of volcanic activity (arbitrary units) that is proportional to the relative amount of material injected into the stratosphere in a 10-year period. Note the periods of major activity in the mid-1600s, and again in the early 1800s, that have been associated with cooler periods.
Source: Prepared by Flohn, based on data gathered by Lamb, 1970.

been given by Lamb (1966), based mainly on botanical evidence, together with rainfall and evaporation data (Table 3.1).

3.3.6
The Period of Instrumental Observations

Quantitative measurements of precipitation were made in Israel over 2000 years ago and in Korea more than 500 years ago. However, comparable climatic records did not exist in Europe before the initiation of scientific societies in Florence (1652), London (1668), Paris (1752), and Mannheim (1783). Since it is virtually impossible to outline here the vast amount of evidence for local and regional climatic changes in the last few centuries—see, for example, von Rudloff (1967) for Europe—only a few general conclusions will be drawn here.

Climatic data are now available from all governmental meteorological services and coordinated by the World Meteorological Organization (WMO). For any investigation of climatic change the observations must be reliable and homogeneous; they should also be representative for a larger area, though in some cases such as rainfall of convective origin this is not usually possible. In most urban sites the temperature records show a local increase of about 1°C compared with values in a rural surrounding, the minimum temperatures rising more than the maximum temperatures. To discuss regional or global variations, such records should be weighted according to the area that they represent. The maintenance of benchmark stations at undisturbed places by meteorological services is of vital importance for surveying large-scale climatic fluctuations. Rainfall records based on area averages should be investigated.

Each comparative investigation of climatic fluctuations shows appreciable horizontal differences, which indicate the dominant role of a redistribution of heat, precipitation, and atmospheric pressure. Such changes do not necessarily affect hemispherical or global averages; they reflect only variations in the atmospheric circulation patterns, which are in themselves an aspect of the climatic fluctuations.

The fluctuations of the atmospheric circulations have been studied most effectively by Lamb (1966) in constructing surface-pressure maps for every January and July since 1790 and estimated 500-mb maps for several periods since the maximum of the last ice age (Lamb, 1968, 1970). From these studies circulation parameters can be derived which indicate surprisingly strong fluctuations, not only of the zonal characteristics but also of the intensity of the meridional flow and of the position of the meridional troughs and ridges of the Atlantic (Figure 3.5).

Of primary importance is the study of global or, at least hemispherical, averages that reflect an imbalance of climatogenetic factors (see Chapter 5). As in the postglacial period discussed earlier, some large-scale variations of temperature occur nearly simultaneously on both hemispheres and thus affect global averages. Unfortunately, hemispherical or global averages have been available only since about 1880 (Figure 3.6). They reveal a slow rise of temperature (of the order of 0.008°C per year) up to about 1945 and then an apparently more rapid drop. The possible effect of human activity on these global averages will be discussed in Part IV. The role of volcanic activity between 1912 (Katmai) and 1947 (Hekla) and the moderate activity since 1956 certainly contribute

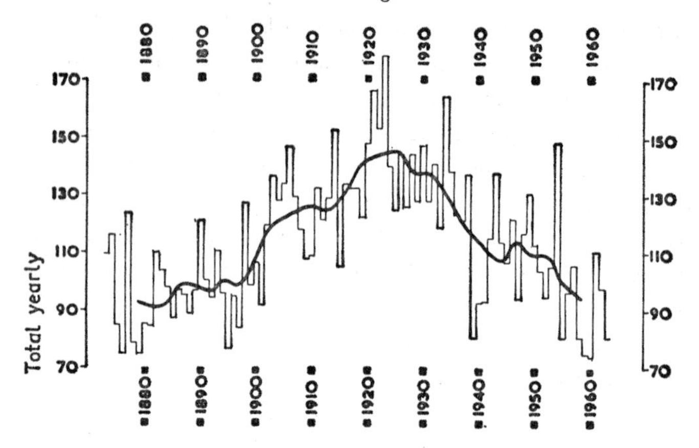

Figure 3.5 Total number of days in the year when there were predominantly west winds ("westerly type" of circulation) over the British Isles, from 1873 to 1963. The solid line shows the course of the 10-year running mean. These changes of circulation pattern can be related roughly to changes in mean temperature and precipitation for the region.
Source: Lamb, 1966.

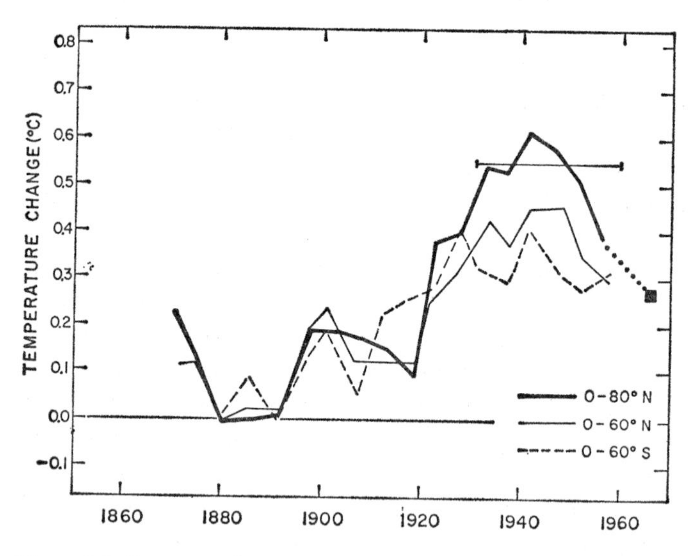

Figure 3.6 Mean annual temperature for various latitude bands, 1870 to 1967. The values up to 1960 were those published ten years ago by Mitchell (1961), and he subsequently (1970) added a value for 1965 to 1967 based on Northern Hemisphere (0° to 80° N) data of Scherhag (1965, 1966, 1967). The horizontal bar shows the mean value of temperature in the 0° to 80° N band for 1931 to 1960.

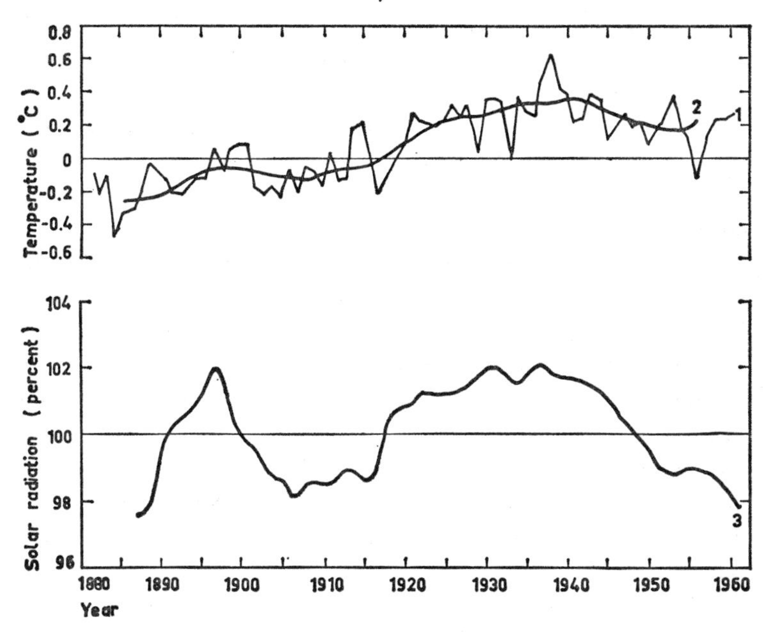

Figure 3.7 A record of mean annual mid-latitude temperatures for the Northern Hemisphere (curve 1) and 10-year running means (curve 2) in the top part, and in the bottom part (curve 3) the mean flux of solar radiation reaching the surface (1886 to 1960) expressed as a percentage of the mean value for the entire period.
Source: Budyko, 1971.

to the observed trends. Figure 3.7 (Budyko, 1971) indicates a rather strong coincidence between hemispherical averages of temperature and solar radiation.

It should be stressed that these climatic fluctuations are usually most important in those areas where conditions are marginal for the support of human life. This is the case in polar regions where climate and its extension of ice greatly affect living conditions, as well as in the transition zone between semihumid and arid climates where the vagaries of rainfall—with more or less constant potential evaporation—strongly affect both agricultural productivity and human economy. In such areas climatic fluctuations with a time scale of only 5 to 10 years are of really vital importance, even if they are considered by some climatologists as only unavoidable statistical noise. For example, the rapid rise of the

Dead Sea around 1895 and the abrupt decline of the Nile runoff at Aswan near 1900 (Lamb, 1966) reflect regional water-balance variations of great economic importance. Historically, the level of the Caspian Sea has undergone especially large fluctuations; however, in the last decades there has been a rapid fall by some 28 meters partly caused by the introduction of artificial lakes and large-scale irrigation in the Volga Basin.

Unfortunately, homogeneity and representativeness of individual rainfull records are not sufficient to study the fluctuations of the water budget, nor are sufficient long-term data of actual (or potential) evaporation available. Because of the large gaps in rainfall measurements, especially over the oceans, it is impossible to investigate possible variations on the global water budget and the factors that influence it. Even long-term global averages cannot be considered as absolutely accurate because of the strong interrelations among the radiation balance, heat, and water budgets.

The most recent development of temperature fluctuations can be shown, using the data for 1961 to 1970 (see Figure 3.8), in com-

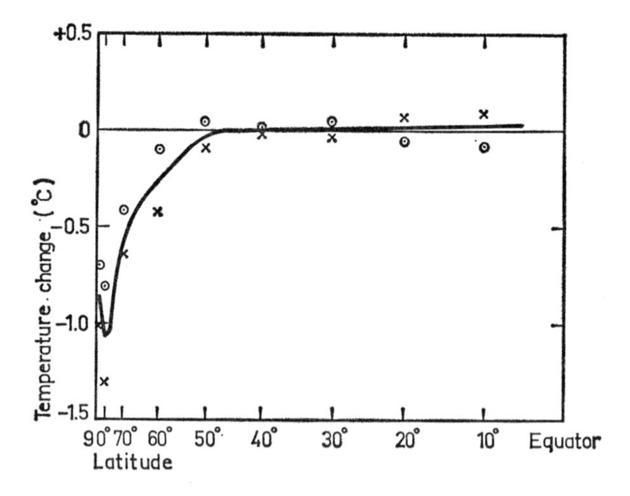

Figure 3.8 Mean temperature in the Northern Hemisphere for 10° latitude belts, expressed as a deviation from the average of 1931 to 1960. The solid line gives the change in the decade 1961 to 1970, the circles the changes 1961 to 1965, and the crosses 1966 to 1970.
Source: Prepared by Flohn using data published by F. R. Pütz, 1971.

parison to the last climatological "normal" (1931 to 1960). While the differences between those of the last decade and the normal are insignificant below latitude 55° N we observe an appreciable cooling in subpolar and polar latitudes, with an average decrease of 0.5°C. This is accentuated in the Atlantic and Siberian sectors of the Arctic where the 10-year average of Spitzbergen is 2°C below normal and the winter temperatures (December–March) of Franz-Joseph have dropped nearly 6°C with little variation during the summer. In 9 of 10 years the arctic regions generally have been colder than before; such low values have not been observed since the beginning of our century.

3.3.7
Recommendation

We recommend that scientists from many disciplines be encouraged to undertake systematic studies of past climates by all available techniques, including studies of historical documents, peat bogs, glaciers, ocean sediments, fossils, and so forth. Information on climates when the Arctic Ocean was free of ice would be particularly valuable in understanding the causes of climate change.

3.4
Processes Governing the State of the Atmosphere

The brief description of the history of climate contained in the foregoing section already suggests some of the major processes or mechanisms of climate changes. We shall next consider more systematically but still in a qualitative fashion the major factors that determine climate, and the peculiar difficulties that are inherent in the formulation of a theory that would explain climatic change, whether natural or man-made.

3.4.1
Distinguishing between Natural and Man-made Effects

The problem of determining whether or not man's activities have had or may have a significant effect on the evolution of global climate is not one of establishing simple cause and effect. It is complicated in two distinct ways. First, it is difficult to predict the quantitative effects of any cause, natural or man-induced. Second, it is even more difficult to distinguish the effects of natural causes from those of human origin.

With regard to the first of these difficulties, the chain of events from original cause to ultimate effect is far too long and intricate to understand in terms of one or a few interacting physical processes. It is sometimes argued, by analogy, that it is not necessary to understand the theory of combustion in order to start a fire by striking a match. In the case of the atmosphere and its underlying surfaces, however, the analogy breaks down. The energy converting processes are so diverse and intertwined that it is not immediately apparent that addition of energy to the total system will make it more unstable or more stable. Such simplistic arguments cannot be applied to systems as complicated as the combined atmosphere, ocean, and earth.

With regard to the second difficulty, it should be reemphasized that the significance of man's intervention in the evolution of global climate must be viewed against the background of "natural" variations of climate, that is, those that would have occurred or would occur if man and all his works were obliterated from the face of the earth. If, for example, the magnitude of man's effects on global climate were negligible in comparison with those of "natural" variations, it is probable that the human species and its ecology would only have to accommodate to situations not very different from those of the past few thousand years. On the other hand, if human and "natural" effects are of comparable magnitude, the problems of explaining either are of comparable difficulty.

We are thus led to the necessity of understanding the causes of all climate change, whether natural or man-induced. With reference to the complexity of the system, what is plainly called for is a general theory of climate and climate change. Only with the formulation and development of such a theory (along sound physical and mathematical lines) will we ever be able to establish the magnitude of man's ultimate impact on global climate.

3.4.2 General Formulation of a Dynamical Theory of Climate

The general aim of a theory of climate is to predict the global or large-scale climate of the atmosphere. We take as given the character of the sun's radiation output; the shape, attitude, and orbit of the earth; the prevailing condition of the oceans and land surfaces; and the composition of the earth's atmosphere.

Although there are certain mathematical peculiarities and difficulties in formulating such a predictive theory, it is fair to say that the physical and mathematical substructure for a general theory of climate is well established and is well known to any college physics student. The fundamental theoretical basis is contained in the laws of conservation of momentum, mass, and energy, combined with thermodynamical and chemical laws governing the change of composition or state.

In order to see more clearly how this theoretical structure applies to the evolution of the coupled system of earth and its atmosphere, let us first examine the dynamical principles that govern the behavior of a passive atmosphere, responding to hypothetically known influences of the earth's land and sea surfaces and to astronomically determined radiation input. Under such hypothetical circumstances, the behavior of the atmosphere would be completely determined, but it would still be extremely complicated. It depends on the vertical exchange of heat, momentum, moisture, and other constituents between the atmosphere and the underlying land or sea surfaces, a process associated with quasi-random, small-scale fluctuating motions and thermal convection near the surface.

In addition, the atmosphere's behavior is directly affected by local heating or cooling, in part due to absorption of direct and reflected shortwave radiation from the sun, absorption of infrared radiation from the surface and other absorbers in the atmosphere, and to emission of infrared radiation. Heating by radiative processes depends, in turn, on the amount of solar radiation impinging on each element of the atmosphere and on the nature and concentration of absorbing and emitting constituents of each element. Thus it is also necessary to specify or predict the distribution of atmospheric constituents—principally ice, liquid water, water vapor, carbon dioxide, and a wide variety of naturally occurring and man-made particles.

The latter fact introduces a new set of complications, since the atmospheric water content is continually changing phase in a manner that depends on the changing motions and local thermodynamic state of the atmosphere. Moreover, the particulate content of the atmosphere is variable and strongly dependent on the strength of sources and sinks, dispersion rates, and rates of chemical reactions that take place within the atmosphere (see Chapter 8).

The other major factor in the heating of the atmosphere is the release of latent heat when water vapor is condensed into liquid water droplets. The rate at which this process operates depends not only on the rate of infrared emission and absorption of solar and infrared radiation but also depends crucially on the rate at which air is cooled adiabatically by lifting of air to lower ambient pressure. It thus depends on the motions of the atmosphere. Moreover, the rate of condensation depends to some extent on the nature and concentration of the small particles on which water vapor condenses and, as in the case of radiative heating, depends on the distribution of atmospheric constituents.

From this brief outline of the major mechanisms of atmospheric heating, it may be appreciated that even the "internal" dynamics of the atmosphere and its processes of change are highly interactive—radiative heating affecting the condensation rate, the existence of clouds of condensed water droplets affecting the reflection and absorption of radiation, and the existence of variable particulate constituents affecting both of the major heating processes.

3.4.3
Variability of Surface Characteristics

In reality, the question is still more complicated, for the hypothetical premises of the foregoing discussion are not realized. The influences of the underlying land or sea surfaces cannot be arbitrarily specified. They are also variable and depend on a new set of interactions between the atmosphere and the land or sea surfaces. The exchanges of heat, momentum, and water vapor between the atmosphere and the surface are governed by the ground temperature, ground or sea roughness, sea temperature, soil moisture, and snow or ice cover.

The surface characteristics are determined, in turn, by the net radiation received or lost, the exchange of heat and water vapor, precipitation, water transport within the surface, and by transport processes internal to the sea and the subsurface ground. The soil temperature, for example, is strongly influenced by the evaporation of soil moisture and conduction of heat in soil of variable heat capacity. Similarly, soil moisture depends on evaporation, precipitation, and diffusion of moisture in soil of variable porosity. Some of these surface characteristics are discussed in more detail in Chapter 7.

The characteristics of the sea surface are also variable and depend strongly on long-term average conditions in the atmosphere. The sea-surface temperature is affected markedly by oceanic currents that are induced by the motion of the atmosphere. It will be apparent, therefore, that the "external" dynamics of the underlying surface and its processes of change are also highly interactive and are strongly linked with the atmosphere's "internal" dynamics.

We shall discuss in greater detail the various radiative, thermodynamic, and transport processes that comprise the "internal" and "external" dynamics of climate in Chapter 5.

References

Additional references for Chapter 3 are on page 294.

Broecker, W. S., and van Donk, J., 1970. Insolation changes, ice volumes and the O^{18} record in deep-sea cores, *Reviews of Geophysics and Space Physics, 8:* 169–198.

Budyko, M. I., 1963. *The Heat Budget of the Earth* (Leningrad: Hydrological Publishing House).

Budyko, M. I., 1971. *Climate and Life* (Leningrad: Hydrological Publishing House).

Budyko, M. I., and Vasischeva, 1971. (In press).

Dansgaard, W., Johnsen, S. J., Clausen, H. B., and Langway, C. C., 1971. *The Late Cenozoic Glacial Ages,* symposium edited by K. K. Turekian (New Haven: Yale University Press).

Emiliani, C., 1955. Pleistocene temperatures, *Journal of Geology, 63:* 538–578.

Emiliani, C., 1966. Isotopic paleotemperatures, *Science, 154:* 851–857.

Ewing, M., and Donn, W. L., 1956. A theory of ice ages, I, *Science, 123:* 1061–1066.

Ewing, M., and Donn, W. L., 1958. A theory of ice ages, II, *Science, 127:* 1159–1162.

Fairbridge, R. W., 1961. Convergence of evidence on climatic change and ice-ages, *Annals of the New York Academy of Science, 95:* 542–579.

Flint, R. F., 1957. *Glacial and Pleistocene Geology* (New York: John Wiley & Sons, Inc.).

Flohn, H., 1964. Grundfragen der Paläoklimatologie im Lichte einer theoretischen Klimatologie, *Geologische Rundschau, 54:* 504–575.

Flohn, H., 1969. Ein geophysikalisches Eiszeit-Modell, *Eiszeitalter und Gegenwart, 20:* 204–231.

Frenzel, B., 1967. *Die Klimaschwankungen des Eiszeitalters* (Braunschweig: F. Vieweg).

Heusser, C. F., 1966. Polar hemispheric correlation: Palynological evidence from Chile and the Pacific northwest of America, Royal Meteorological Society, World Climate from 8000 to 0 B.C. (London: Proceedings of the International Symposium held at Imperial College, April 18 and 19, 1966).

Hoinkes, H., 1968. Wir leben in einer Eiszeit, *Umschau, 68:* 810–815.

Johnsen, S. J., Dansgaard, W., and Clausen, H. B., 1970. Climate oscillations 1200–2080 A.D., *Nature, 227:* 482–483.

Ku, T. L., and Broecker, W. S., 1967. Rates of sedimentation in the Arctic Ocean, edited by M. Sears, *Progress in Oceanography, 4:* 95–104.

Kutzbach, J. E., Bryson, R. A., and Shen, W. C., 1968. An evaluation of the thermal rossby number in the Pleistocene, American Meteorological Society, *Meteorological Monographs, 8:* 134–138.

Lamb, H. H., 1966. *The Changing Climate* (London: Methuen and Company), p. 236.

Lamb, H. H., 1968.

Lamb, H. H., 1970. Volcanic dust in the atmosphere: with a chronology and an assessment of its meteorological significance, *Philosophical Transactions of the Royal Society, 266:* 425–533.

Milankowitch, N., 1930. Mathematische Klimalehre und astronomische Theorie der Klimaschwankungen, *Handbuch der Klimatologie,* by W. Köppen and R. Geiger, *1A* (Berlin).

Mitchell, J. M., Jr., 1961. Changes of mean temperature since 1870, *Annals of the New York Academy of Sciences, 95:* 235–250.

Pütz, F. R., 1971. Klimatologische wetterkarten der nordhemisphére für dul temperature im dezen niusn 1961/70, *Bilage zur Berliner Wetterkarte, 106:* 23.

Schwersbach, M., 1961. *Das Klima der Vorzeit* (Stuttgart: F. Enke), 2. Auflage, p. 275.

Shaw, D. M., and Donn, W. L., 1968. Milankovitch radiation variation: a quantitative evaluation, *Science, 162:* 1270–1272.

Shepard, F. P., and Curray, J. R., 1967. Carbon-14 determination of sea level changes in stable areas, *Progress in Oceanography, 4:* 283–290.

Scherhag, R., 1965, 1966, 1967. Temperature data published in *Berliner Wetterkarte,* The University of Berlin (annual series).

von Rudloff, W., 1949. Climate of the humid tropics, *Der Wetterlotse, 19:* 89–95.

Willett, H. C., 1949. Long-period fluctuations of the general circulation of the atmosphere, *Journal of Meteorology, 6:* 34–50.

Wilson, A. T., 1964. Origin of ice ages: an ice shelf theory for pleistocene glaciations, *Nature, 201:* 477–478.

Woldstedt, P., 1958, 1961, 1965. *Das Eiszeitalter, Grundlinien einer Geologie des Quartärs,* Vol. 1, third edition, Vols. 2 and 3, second edition (Stuttgart: F. Enke).

4 Man's Activities Influencing Climate

4.1 On Predicting the Future

There can be little doubt that man, in the process of reshaping his environment in many ways, has changed the climate of large regions of the earth, and he has probably had some influence on global climate as well—exactly how much influence we do not know. In this chapter we shall identify those ways in which he has had an influence, or could have, and note briefly some of the trends. These trends obviously have implications for the future which will be discussed later in Part IV.

In speaking about the future of man and his climate we are faced with two important fields for consideration, *meteorological* (or geophysical) and *social*. Whereas we do have a rudimentary theory of climate and a good deal of knowledge about how things behave in the physical world, when we turn to a forecast of man's behavior we must resort to simple extrapolation of present trends, with only slight shadings to take account of the more obvious interactions that we can foresee. There are no laws that we know of, no mathematical "models," that will allow us to predict the future course of human affairs better than such extrapolations. Therefore, in the end we can forecast only what *could* happen *if* mankind proceeds to act in a certain way, more or less as he is acting now.

4.2 Atmospheric Contamination

One of the most clearly evident influences of man on his environment is his direct contamination of the atmosphere. In cities air pollution is sometimes acute, but even in the most remote places in the world it is possible to detect traces of man-made contaminants in the air.

Particles and certain trace gases influence the heat balance of the atmosphere because they alter the flux of solar and infrared radiation. Particles are also involved in the initiation of condensation and freezing in clouds. In this section we shall deal in rather general terms with the factors that govern the distribution of atmospheric contaminants, leaving a more detailed treatment for Chapter 8.

4.2.1
Zonal Distribution of Particles and Their Sources

Since particles and their gaseous precursors (sulfur dioxide, for example, that turns to sulfate) have a mean residence time of less than one week in the lower atmosphere before being washed out, their concentration decreases downwind from the source. Therefore, the distribution of sources and the prevailing wind will determine the concentration of airborne particles to a large extent.

It is important to recognize that the human activities that produce particles are far from uniformly distributed over the earth but are concentrated in latitudinal belts. This zonal distribution becomes particularly important, since the winds themselves tend to be zonal in character.

The chief wind systems of the world in the lower atmosphere are the easterly trades of the subtropics that are notable for their steadiness, the westerlies of the temperate zones that are variable due to migrating high- and low-pressure systems, and the polar easterlies that are relatively weak and variable. The continental monsoons, which are most well developed over Asia, are wind systems determined by continent-ocean temperature contrast and therefore reverse themselves from summer to winter.

In spite of the variability of the winds, experience with tracers such as balloons and radioactive particles shows that contaminants will indeed move zonally for the most part with a more gradual spreading to the north and to the south due to meandering of the flow patterns, on which is superimposed a slow meridional drift that depends on latitude and season. The same is more or less true at all levels, including the stratosphere.

The primary continental contributions to atmospheric dust are soil and rock debris from the arid regions of the world. Since these regions have their greatest expanses in the mid-latitudes of the Northern Hemisphere, larger amounts of materials are mobilized above the earth's surface in such regions. In addition, the deserts in northern Africa and Southwest and Central Asia provide substantial amounts of continental debris for transport in the continental monsoons and trade winds. As will be shown in Section 4.4.2, man has played a part in creating this source of dust.

The impact of such sources of particles can be readily seen in the composition of deep-sea sediments in the same latitudes as the source areas. For example, in the North Pacific there are bands of clay mineral illite and quartz whose origin appears to be the desert and steppe zones of Europe and Asia (Griffin, Windom, and Goldberg, 1968). A record of the transport of lateritic material from the western part of Australia to the Indian Ocean by the prevailing southeasterly trades is found in the Indian Ocean sediments west of Australia.

Volcanoes eject their materials into the atmosphere along two primary latitudinal belts: one narrow zone just below the Arctic Circle, 56° N to 65° N, and including the volcanoes of Kamchatka, the Aleutian Islands, Alaska, and Iceland; and the other broadly centered about the equator, 8° S to 15° N, and including the volcanoes of Indonesia, the Philippines, Central America, Ecuador, the Galapagos, and the Caribbean (Cronin, 1971). During the past decade there have been seven major eruptions powerful enough to inject material into the stratosphere, four in the Arctic Circle belt and three in the equatorial zone. The dusts from the northern area have never been reported to reach the Southern Hemisphere, although volcanic dust released at the equator is observed in both the Northern and Southern Hemispheres. As a consequence of volcanic activity, then, the Northern Hemispheric stratosphere becomes more turbid than its southern counterpart (Lamb, 1970).

The dominant use of energy and of material by man takes place in the mid-latitudes of the Northern Hemisphere. The leaks of gases and particles to the atmosphere, whether deliberate or unintentional, constitute contamination. The gross national product (GNP) of a country is related to the flow of materials through it and to the utilization of energy and can be used to give a measure of potential atmospheric contamination. Nine countries (United States, Soviet Union, Japan, Federal Republic of Germany, France, Great Britain, Canada, China, and Italy) have GNPs greater than $\$7 \times 10^{10}$ and a combined value of $\$1.9 \times 10^{12}$.

All are at mid-latitudes in the Northern Hemisphere. The next twelve countries have GNPs above $\$1.5 \times 10^{10}$ (India, Poland, Australia, Sweden, Netherlands, Spain, Democratic Republic of Germany, Brazil, Mexico, Belgium, Switzerland, and Czechoslovakia) with a total of $\$2.0 \times 10^{11}$. Of these, India and Brazil are in the tropics, and only Australia is in the Southern Hemisphere. Thus, as a first approximation, it is clear that man's atmospheric injections from energy use and industrialization take place under a single transporting agency, the prevailing westerly flow in the Northern Hemisphere.

4.2.2
Gaseous Contaminants

Where man injects particles into the atmosphere he usually releases gaseous contaminants also. Some of these are chemically stable gases, such as carbon dioxide, hydrogen, freon, sulfur hexafluoride, and methane and, therefore, diffuse very widely to add to the total trace-gas component of the atmosphere. Others are not so long-lived in the atmosphere and are converted to other gases or become transformed into particles. Examples of such short-lived gases or volatile substances are sulfur dioxide, hydrogen sulfide, ammonia, unburned hydrocarbon fuel, polychlorinated biphenyls, DDT, and so forth. Since they have a limited lifetime in the atmosphere they will, like particles, tend to remain in the mid-latitude belts where they are mostly produced.

In order to understand the full impact of the short-lived gaseous or volatile substances, we must consider them as being transported by the atmosphere from their sources in industrial areas, along with particles that are emitted by the same sources, and then deposited in sites where they can become part of the hydrosphere and biosphere. The oceans make up the largest part of the biosphere, and indeed this is where a majority of such substances eventually deposit because there is a continuous transport of man's effluents into the marine environment. Freshwater lakes also receive appreciable amounts of wastes, as evidenced by the growing acidity of rain in much of Europe and the effect of this on certain Swedish lakes (Lundholm, 1970).

4.3 Energy Production and Release

Since the release of particles and gases into the atmosphere is commonly called "air pollution," the similar release of heat has often been termed "thermal pollution." In rivers and estuaries it has become something of an issue because of its effect on plants and animals in the water, and it now appears that, with our escalating demand for more and more power, we may have to consider seriously its influence on the atmosphere as well.

The human activity that contributes the most additional energy is the burning of fossil fuels: coal, gas, and petroleum products. The rapidly increasing contribution of nuclear energy for electrical generation and desalinization of water (now at 0.2 percent of the total) must also be included in any future projections.

Other energy sources that have been tapped are part of the natural energy processes of the earth-atmosphere system. Hydroelectricity, which now contributes about 2 percent of global energy production, uses only the potential (converted to kinetic) energy of water stored by the hydrological cycle in mountains. The very small contributions of geothermal, direct solar, and wind energy need not be considered separately. Unfortunately, the existing data do not allow us to rank these different sources.

4.3.1 Distribution of Artificial Sources of Heat

The global energy production as of 1970 is estimated to be 7.5×10^9 tons/year coal equivalent (OECD, 1969; *The Times*, 1968). Using 8100 cal/g coal as the energy equivalent,* the global energy production can be calculated to be 6×10^{19} cal/year, or 8×10^6 megawatts (MW) for 1970.† (We will treat the growth rate of energy production in the next section.) Table 4.1 represents various releases of energy in comparison to net surface radiation from the sun.

Early estimates for particular places have been given by W.

* This assumes complete combustion, and is therefore a maximum estimate.

† This is considered an "upper-limit" figure in comparison to the SCEP (1970) estimate of 5.5×10^6 megawatts (source: United Nations, 1968).

Table 4.1
Sources and Magnitude of Certain Energy Releases

Description of Energy Release	Magnitude (W/m^2)
Net solar radiation at earth's surface (dependent on latitude)	approx. 100[a]
Urban industrial area estimate[b]	12
1970 energy production distributed evenly over all continents	54×10^{-3}[c]
1970 energy production distributed evenly over the whole globe	15.7×10^{-3}[d]
Annual net photosynthetic production of the continental vegetation cover	0.13[e]
Global average heat flux from the earth's interior	0.062[f]

[a] Budyko, 1963.
[b] Assuming 75 percent of the energy consumption to be concentrated in a total industrial area of the globe of 0.5×10^6 km^2.
[c] This figure was obtained using the 1970 energy production value of 8×10^6 megawatts, discussed in Section 4.3.1.
[d] Same as c.
[e] Lieth, 1964.
[f] Higher locally by a factor of 10 to 50 in geothermal areas.

Schmidt (1917) for Berlin and Vienna. However, these assumed incomplete combustion and thus underestimated energy production by 50 percent. A more recent estimate has been given for Sheffield in the year 1952 (refer to Table 4.2). In highly industrialized countries the average energy consumption per head E_h is rather uniformly distributed; it varies mostly between 5 and 12 kilowatts (kW) per capita. In partly rural countries with a warmer climate (Japan, Italy) it is lower but may reach more than 20 kW per capita in cities with severe winters. Such country-averaged figures allow a minimum estimate for cities and industrial areas, when no local data are available. If E_h is given in kW per capita (or head) and the population density P in heads/km^2, the energy flux density is given by

$$F = 10^{-3} E_h \times P \text{ W/m}^2$$

This relation permits us to draw maps of the energy flux density. Figure 4.1 shows the relationship between E_h and P (on a logarithmic scale) for several cities, industrial areas, and large countries.

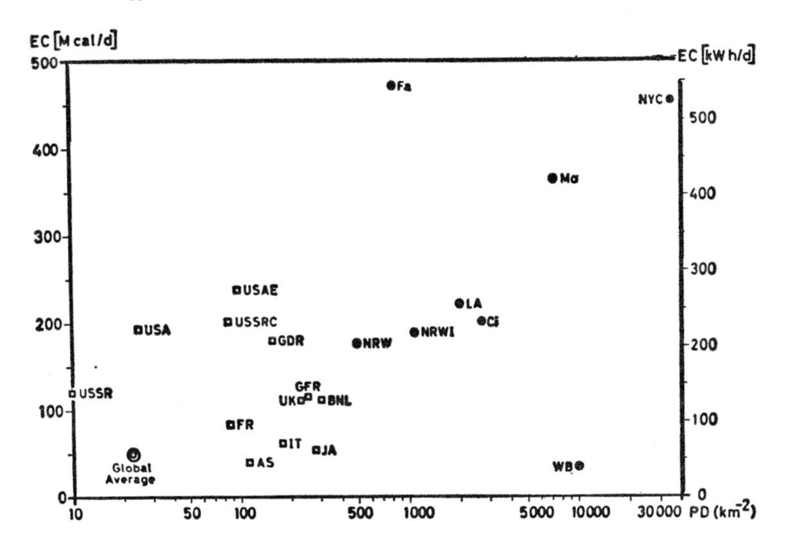

Figure 4.1 Energy consumption (EC) per capita and population density (PD). Source: Data from Tables 4.2 and 4.3.

From W. Schmidt's data (corrected for complete combustion) we obtain, at the beginning of this century, for Vienna 0.75 kW and for Berlin City 1.4 kW per capita. The physiological energy production of man is only about 0.12 kW and can frequently be neglected, as well as the metabolism of animals.

Table 4.2 gives various data, on the local scale, for industrial and urban areas, mostly valid for the years 1965 to 1968. Data related to the building area may overestimate the energy flux concentration. Other data refer to administrative units, including many uninhabited areas such as forests, lakes, mountains, and so forth, where the energy flux density tends to be underestimated. Since in American cities mobile sources consume as much as 25 percent of the total energy, a distribution over suburban areas appears to be justified. Table 4.2 indicates that at the local scale (up to 10^3 km^2), the artificial energy flux density is of the same order of magnitude as is the natural net radiation, and in very heavily industrialized cities, such as Moscow, it is even larger (Lydolph, 1970).

Any system for averaging of the artificial heat input into the

Table 4.2
Energy Consumption (EC) Density, Industrial and Urban Areas

	Area (km^2)	Population 10^6	EC density (W/m^2)	EC per Capita, E_h	Average net radiation (W/m^2)
Nordrhein-Westfalen	34,039	16.84	4.2	8.0	50
Same, industrial area only	10,296	11.27	10.2	8.9[b]	51
West Berlin	234[a]	2.3	21.3	2.0	57
Moscow	878	6.42	127	16.8[b]	42
Sheffield (1952)	48[a]	0.5	19	1.6	46
Hamburg	747	1.83	12.6[a]	5.0	55
Cincinnati	200[a]	0.54	26	9.3	99
Los Angeles County	10,000	7.0	7.5	10.3	108
Los Angeles	3,500[a]	7.0	21	10.3	108
New York, Manhattan	59	1.7	630	21.0	93
21 metropolitan areas (Washington-Boston)	87,000	33	4.4	11.2[c]	~90
Fairbanks, Alaska[a]	37	0.03	18.5	21.8	18

Note: The data for Montreal (Oke, 1969), showing 58 W/m^2 during summer, 156 W/m^2 during winter, are in agreement. Hannell (1969) has estimated that the energy output of a steel mill at Hamilton, Ontario, is 385 to 580 W/m^2 (sensible heat) and 48 W/m^2 (latent heat), equivalent to a total of about 430 to 630 W/m^2.
Source: Compiled by H. Flohn from numerous published and unpublished sources.
[a] Building area only. [b] Related to industrial production. [c] Eastern United States.

earth-atmosphere system over large inhomogeneous areas will give only very approximate results. Such values for several countries (Table 4.3) are given here for the sake of comparison. Only in some areas, such as the "Megalopolis" between Boston, Massachusetts, and Washington, D.C., or the industrial area of northwest Germany and southern Belgium between Dortmund and Calais, are the stationary heat sources so densely distributed that area averages for 10^4 to 10^5 km^2 are meaningful (or soon will become

Table 4.3
Energy Consumption (EC) Density for Selected Large Territories

	Area (10^3 km^2)	Population (10^6)	EC MW $\times 10^6$	EC Density (1967) (W/m^2)	EC per Capita, E_h (kW)
Germany, Federal Republic	246	62	0.336	1.36	5.4
Germany, Democratic Republic	108	17	0.150	0.91	5.8
Benelux	73	22	0.124	1.66	5.5
United Kingdom	242	55	0.295	1.21	5.4
France	573	50	0.188	0.32	3.8
Italy	299	53	0.160	0.53	3.0
Austria, Switzerland	124	14	0.029	0.23	2.0
Central Western Europe	1,665	273	1.120	0.74	4.5
United States	7,760	196	1.586	0.24	9.3
United States, 14 eastern states	932	90	1.040	1.11	11.6
U.S.S.R.	22,400	233	1.380	0.05	4.4
Central Russia	256	22	0.219	0.85	9.8
Ukraine	604	47	0.305	0.50	6.6
Japan	366	99	0.263	0.71	2.7

Sources: Lydolph, 1970; OECD, 1969 (unpublished).

so). In such areas 5 to 10 W/m^2, or about 10 percent of the net solar radiation, is already reached.

4.3.2
Growth Rate of Energy Production

The annual growth rate of energy production is gradually rising. Table 1.3 of the SCEP Report shows 5.7 percent per year for the world, based on United Nations data. In some countries (Italy, Japan) the annual growth rate already reaches 10 percent and more. (An annual growth rate of 5.5 percent is equivalent to a rise by a factor of 5 in 30 years.) Thus, we should expect a further extension of industrialized areas with a significant contribution of artificial energy, and there will be many areas of 10^3 to 10^4 km^2 where the additional input of energy is of the same order of magnitude as the natural net radiation. On the global scale, the present insignificant contribution of 0.05 W/m^2 to the continental surface heat budget would rise to the 1 percent level of the continental net radiation average of 67 W/m^2 after about 40 years, and to the 10 percent level within 100 years if the rate of growth continues. A similar estimate based on a growth rate of only 4 percent has been made by Budyko, Drosdov, and Judin (1966). With an assumed growth rate of 5.5 percent in western and central Europe (1.56×10^6 km^2), the artificial heat input will reach about 3.8 W/m^2 by 2000 A.D. and nearly 50 W/m^2 at 2050 A.D.; similar figures can be obtained for the eastern United States. However, our assumptions of a constant energy growth rate and a uniform distribution of artificial heat inputs above larger areas may very well be unrealistic.

4.4 Land-Surface Alterations

Man changes the landscape in many ways. Furthermore, he has been doing this for thousands of years. For example, analyses of cores in European peat bogs reveal that there was a decline in tree pollens after about 3000 B.C., associated with the creation of a shepherd-farmer culture and a consequent reduction in forest cover (Seddon, 1967). Throughout the subsequent centuries, man's impact on the environment has been sometimes for the better (draining of swamps, for example) and sometimes for the worse (scarring the countryside by "urban sprawl" and creating semi-deserts in arid regions by overgrazing animals).

Few land areas are today in unaltered or "base-line" states, and the rate of man-made change is increasing. This is to be seen, for example, in the rapid expansion of built-up areas and the clearing of jungles and swamps. In addition, twentieth-century technology allows man to penetrate into remote areas in search of resources and amenities. Mining in the sub-Arctic and drilling for oil and gas in the Arctic and beneath the oceans are now feasible; and the aircraft, the jeep, and the snowmobile provide great mobility in penetrating the wilderness for pleasure. Even in the central oceans man leaves his signature in the form of oil slicks. In some cases, of course, the trend is being reversed in a small way, and marginal farmland is being returned to a forest cover in various parts of the world.

4.4.1
Urbanization and Industrialization

The industrial and urban activities within cities and in the areas between them alter the landscape as well as injecting materials and heat into the atmosphere and adjacent water bodies. In principle, all of these changes can influence the parameters determining climate. Most of the man-made changes in urban land use, however, are especially difficult to quantify at the present time.

The impact of industrialization and urbanization can be categorized in the following way:

Material injections—water vapor, volatile organic compounds, and particles (see Sections 4.2, 5.1, 5.2, and Chapter 8).

Heat injection (see Section 4.3 and Chapter 7).

Surface albedo changes—buildings and roads with different reflecting properties (see Chapter 7).

Changes in roughness—buildings (see Chapter 7).

Any process that changes the heat balance of the atmosphere can influence climate, at least locally, and the first three of the preceding effects do just that. However, the albedo change is generally not very great except when there is a snow cover and the cleared streets absorb more sunlight. The greater heat experienced in a city on a sunny afternoon, compared to the surrounding countryside, is due mostly to the small heat capacity of the walls and roofs of the buildings and the absence of any cooling caused by evaporation from vegetation. The last-listed effect also plays a role here, because there is less ventilation between buildings than in the open country, and the air at street level therefore remains relatively stagnant. However, this effect probably does not play a major part in the modification of the climate of the region as a whole.

In spite of our uncertainties about the quantitative influence of cities on the climate, we are concerned about the rapidly increasing conversion from rural to urban land use. In the Lake Erie basin, for example, the projections are that within 50 years the entire area will be withdrawn from the agricultural land bank. The suburban sprawl is a widespread phenomenon of the twentieth century, and we urge that regular land-use inventories be undertaken and coordinated internationally in order to be able to assess the influence this trend will have in the future.

4.4.2
Nonurban Land Use

The most widespread modification of the climate by man in the past has been inadvertently achieved by the conversion of the natural vegetation into arable land and pastures. During the last 8000 years increasing areas have been converted from steppe and forested steppe into arable land, where the soil remains more or less bare during several months while it is being cultivated. During this season the CO_2 production of soil bacteria continues, while the immediate consumption for assimilation ceases.

The conversion of well-developed forests began somewhat later—in large portions of Europe, the eastern portion of the United States, and the mountains of Southwest Asia from Turkey to Afghanistan, all of which were originally covered by dense coniferous or deciduous forests.

In tropical latitudes the savannah grasslands are entirely manmade, the natural type of vegetation in the semihumid tropics being a dry deciduous forest that has been gradually destroyed by the widespread bush fires ignited by local inhabitants during the dry seasons. In these areas natural forest fires could have been caused also by lightning during the rainy seasons, but only comparatively rarely (perhaps once every 10 to 50 years) and having short durations because the fires would not burn so extensively over the damp vegetation. The optical effects of particles entering the atmosphere from these still common bush fires are discussed in Chapter 8.

This type of conversion is widespread in tropical Africa, in northeastern Brazil, and in some semihumid areas of Central America and Southeast Asia, while in India the conversion into arable land has predominated; it covers at least 10^7 km². A rough estimate of the conversion of forests and forest-steppe in temperate latitudes is 1.5 to 2×10^7 km². The net result is that the land use of 2.5 to 3×10^7 km², or 18 to 20 percent of the total area of the continents, has been drastically changed, with substantial consequences for the climatological heat and water budget. Unfortunately more precise figures are not available.

In many arid or semiarid areas the increasing water demand has led to a gradual shrinkage of groundwater resources. If this shrinkage continues for several decades, it will result in an imbalance of the regional water budget (discussed further in Sections 4.5 and 7.4). Furthermore, the natural vegetation is rapidly being lost over larger and larger areas, a consequence of an uncontrolled increase of the number of domestic animals (overgrazing), especially in Africa and Southwest Asia. Where the original state of vegetation has been retained, for example, in military compounds, the contrast is dramatic. In such a case, in southern Tunisia near Nefta, the coverage of the vegetation inside of an area fenced 60 years ago is 85 percent, in contrast to 5 percent outside of it; here the original aspect of a dry steppe has changed into a semidesert, without any appreciable recent variation of precipitation (about 80 mm/yr). Similar observations have been made by Bryson (1967) near Jodhpur, India. In many areas of the Old World the deserts apparently are spreading, with an apparent speed of about one or more kilometers per year, depending on the density of the population, as a consequence of grazing animals (notably goats).

Concern has been expressed concerning the removal of forests in the tropics where there is considerable rainfall and the vegetation depends for its existence on retaining the essential minerals in a relatively thin layer of humus. The subsoils have been leached of their nutrients by centuries of heavy rainfall. In such areas vast expanses of tropical forests have been transformed into desertlike surfaces of laterite cements through attempts to cultivate food crops. After one or two growing seasons the highly weathered soils (devoid of their ground cover and with the humus layer washed away) become barren and covered with impervious, solid layers of kaolinitic and montmorillonitic clays.

4.5
Manipulation of Surface and Underground Water

4.5.1
Damming Rivers to Create New Lakes

Management of river discharge for irrigation, flood control, and the draining of swamps was one of man's earliest organized activities. Recently the establishment of large hydroelectrical power plants has required large-scale damming of rivers with the consequent formation of new lakes, some of them large enough to appear on maps of the world. River flows have become so regulated that in many cases large natural annual fluctuations in their lower parts have been reduced to a very small fraction of the original amplitudes. Rivers have even been diverted across watersheds so that they discharge into widely different parts of the ocean.

All of this has its influence upon the atmosphere. Open water has an albedo substantially lower than that of other surfaces. Frozen, snowcovered water has a very high albedo. Both the amount and the seasonal distribution of absorbed solar heat can thus be changed by altering the area covered by water.

Over land, almost all the received solar energy is given up within at most a few days by longwave radiation, sensible heat transfer to the atmosphere, or by evaporation, because the energy storage of the land is small. Water, however, is capable of storing large quantities of heat that may be released much later—months or in the case of the ocean, even years later. Furthermore, since the water surface may have a temperature quite different from that of land at the same location, the distribution of heat among longwave radiation, sensible heat, and evaporation may be significantly different when a new body of water is created. This is particularly true when a water body occurs in an otherwise arid area (for example, the Great Salt Lake in western United States and Lake Nasser in Egypt) where hot, dry winds from the surrounding land may cause large amounts of evaporation from the water. In this case heat absorbed upon the surrounding dry area is indirectly used to increase evaporation. Downwind the water vapor content of the air is increased, creating effects on the radiation balance, and eventually it can influence precipitation as well (see Section 7.4.3).

4.5.2
Irrigation of Arid Land

Another process modifying climate is irrigation. The 1700 km^3/yr of nonreturned irrigation water (Lvovich, 1969) represents a change of about 5 percent in the runoff of the land areas. The 1700 km^3/yr additional evaporation is approximately 2 percent of the total annual evaporation of land areas.

Although the effects produced are quite different, some idea of the magnitude involved can be obtained by comparing the energy required to evaporate this irrigation water with the energy introduced into his environment by man's use of power. To evaporate 1700 km^3 of water requires over 10^8 megawatts. As is stated in Section 4.3.1 on energy production, man's power production now amounts to about 8×10^6 megawatts. No inferences about the relative importance to climate of irrigation and power production can be drawn from this order-of-magnitude difference, for while energy production enters directly into the overall heat budget of the earth, the direct influence of the energy used for the evaporation of irrigation water is to speed the hydrological cycle. All of the heat used in evaporation is returned to the atmosphere where condensation occurs. Apart from local effects, which are large, the global climatological impact will depend upon such indirect influences as changes in cloud cover and resulting effects on the radiation balance. The estimation of such effects is an extremely complex problem, and, as is pointed out elsewhere in this document, we do not yet have any reliable method for making quantitative calculations of them.

One indirect effect can, however, be calculated, and it turns out to be quite significant. An irrigated area grows vegetation that has an albedo significantly lower than the ground cover that has been replaced. The interesting consequence of this is that while the *local* temperature is lowered by irrigation because of the increased evaporation, *global* temperature is raised because of the decreased reflection of incoming solar radiation. Budyko's (1971) calculation that present-day irrigation leads to an increase of the earth's mean surface temperature amounting to 0.07°C is derived from consideration of this effect.

According to the estimate of Lvovich (1969), by the year 2000 A.D. the water demand for irrigation will nearly double, and the combined demand for irrigation, industry, energy production, and waterworks will be five times as great. Thus, the amount of evaporating water will be three times higher than in 1965. The total effect of irrigation on global albedo and the effects of man's activities on the hydrological cycle must therefore be expected to increase substantially (see Section 7.4).

4.5.3 Influence of Irrigation on Underground Water

In Section 4.4.2 we discussed the changes in natural vegetation that man has brought about in semiarid areas, and there is an important by-product of this that has not received the attention it deserves. This is the effect of "mining" underground water, termed "fossil water," that was accumulated in earlier climatic eras when there was more precipitation than now.

An important example of this is to be found in North Africa. In the classical Saharan oases, originating in a pre-Roman epoch, the recent extension of agriculture, mainly the increase of the number of high-valued (taxable) date palms, was possible only by tapping the subterranean reservoirs of fossil groundwater preserved to a large extent from the last ice age, with a radiocarbon age of 20,000 to 25,000 years (Knetsch et al., 1962). In the oases of southern Tunisia, agricultural production has been raised by at least 50 percent and more probably by 80 to 100 percent. For example, in the Nefzaoua group of oases, a large number of taxable date palms have been planted since 1947. By assuming a more or less linear correlation between agricultural production and water demand, the present water supply can be estimated to be (at least) 50 percent higher than at the end of the last century. This means an increase of the average evaporation by nearly 2 percent in an area of about 35,000 km^2, of which only 150 km^2 are covered by oases. Because the average precipitation in the area has been constant since the beginning of regular observations (about 1903), the observed drop of several meters in the groundwater level can only be interpreted as a consequence of an increasing imbalance of the actual water budget caused by the exploitation of fossil water reserves at a faster rate than can be replenished under present climatic conditons.

While this depletion of an essentially irreplaceable source of water is obviously of great importance to the areas involved, underground water is being removed fast enough in many other places so that the water tables are falling. Groundwater stored in the pores and fractures of soil and rock constitutes about 5 percent of total free water, compared with 93 percent for the ocean and 2 percent for polar ice (Lvovich, 1970). As man has drained it deliberately by pumping or by inducing artesian flows, usually for irrigation, the size of this store is decreasing, and where there is underground salt water along a coast it is moving in to replace the fresh water. This is another case where man is changing the natural balance of the earth (though not necessarily the climate), and it is discussed further in Sections 5.4 and 7.4.5.

4.5.4
Modification of Ocean Waters

The oceans constitute a mass that is several hundred times that of the atmosphere, and it would therefore obviously be much more difficult to change its overall properties in depth. However, the surface layers of the ocean can be changed more easily, and these determine to a large extent the direct interactions between ocean and atmosphere that are such an important aspect of the system (see Sections 3.4.3, 6.8.2, and 7.3).

There is no dearth of proposals for modification of the ocean. For example, over most of the ocean, biological productivity in the upper layers is influenced by the availability of nutrients such as nitrate and phosphate. These nutrients exist in plentiful supply in the water only a few hundred meters beneath the surface. Fertilization could be effected by pumping (perhaps using wave energy) the deeper water to the surface. This deeper water would be considerably lower in temperature than the existing surface water, and the ocean's influence on the atmosphere would change. If, on the other hand, the upwelling is produced by using deep, cold water as cooling water for large power plants, the local surface temperature might be little affected, but oceanic heat transport would be increased, with resulting climatic effects also (see Section 5.3).

Another proposal involves supplementing the freshwater resources of some land areas in the Southern Hemisphere by towing large volumes of antarctic ice to more northerly waters. As a side effect, this would enhance the meridional heat transport now effected naturally by the atmosphere and the ocean.

Although such direct intervention remains at the conceptual stage, man is already having some impact on the ocean as a byproduct of some of his other activities. The diversion of water from one watershed to another distorts salinity patterns in the ocean. A probably minor but well-known example can be found in the Chicago Ship Canal, which diverts water from Lake Michigan (part of the St. Lawrence River drainage system) to the Mississippi system. This diversion leads to an increase in the salinity of the Gulf of St. Lawrence and a decrease in that of the Gulf of Mexico.

Salinity changes induced in this way, and also by regulating river flow rates, produce some rather subtle effects. Low salinity in the upper water increases the stability of this water (fresh water being less dense than salt water) and tends to inhibit downward mixing and to produce a shallower upper layer with less thermal "inertia." The resulting shorter thermal time constant makes the water less efficient in the transport of heat on such long meridional journeys as that of the Gulf Stream water into the Norwegian Sea.

Perhaps more important is the control of river discharge into ocean areas subject to winter freezing. The rate of freezing, the nature of the ice, and the amount of ice formed in the course of a winter all depend upon the salinity and stability of the upper 10 meters or so of relatively fresh water.

4.6
Weather Modification

Most of the effects induced by man on climate are inadvertent byproducts of attempts to attain a variety of goals not associated with either climate or weather. In seeding clouds with freezing nuclei such as silver iodide crystals, however, the objective is to change the weather. Over a period of time this would change the climate.

To some extent, then, the climatic influence of this activity would be a measure of its success. There is not general agreement on how success has been achieved in attempts to modify the weather. There is no doubt that seeding can modify clouds. Where the intent is to increase precipitation, freezing nuclei have been added, with the expectation that supercooled clouds will thereby precipitate more readily. In other experiments common salt particles have been used in attempts to modify warm clouds. Controlled experiments with individual clouds have often given positive results; area experiments in some regions appear to have produced marginal increases in precipitation of the order of 10 to 20 percent at an acceptable level of statistical significance, but in other places no significant increase and even decreases have been reported (U.S. National Academy of Sciences, 1966).

Much of the uncertainty surrounding the effectiveness of cloud seeding arises from the difficulty in obtaining a sufficient body of well-controlled data adequate for statistical tests. The great variability of natural rainfall makes this difficult even in the best-designed experiments.

Whether or not successful rainmaking leads to an increase in global rainfall or merely to its redistribution depends upon a number of factors. It is shown in Table 7.6 that as much as three-fourths to nine-tenths of irrigation water is vaporized by evapotranspiration. If artificially induced rain also largely evaporates, the result is an intensification of the exchange rate in the hydrological cycle. It is then not necessary to "rob Peter in order to pay Paul." On the other hand, if much of the induced rain merely contributes to the runoff, the main result may be a redistribution rather than a net increase in rainfall.

Cloud seeding is employed for purposes other than for rainmaking. In several countries sizable seeding operations are conducted for hail suppression (Sulakvelidze, 1968). Large quantities of seeding particles are also used in experimental attempts to control tropical hurricanes. If this work proves successful, it will offer man the opportunity to exert a very considerable influence on climate. Hurricanes do more than cause coastal flooding and collapse buildings. They can contribute substantially to needed precipitation in some areas and can influence evaporation, ocean mixing, and atmospheric heat transfer. If control becomes possible, it will thus have to be exercised with great judiciousness and with consideration for more than local and immediate effects.

Like so many human endeavors, cloud seeding is showing evidence of unexpected side effects. Several authors, including Elliot and Grant (unpublished), report substantial "downwind" effects that indicate influence of seeding operations well beyond the "target" area. Such effects require the presence of some re-enforcing mechanisms that are not now understood.

The possibility of such unexpected long-range effects demands much more understanding and control over seeding efforts than now exists. At the very least, information on the timing and nature of seeding operations should be centrally collected and made available to the scientific community. Without such information it is impossible to ascertain whether individual seeding experiments are being influenced, perhaps invalidated, by other seeding elsewhere, and such information is even more important in assessing regional effects.

4.7 Transportation

4.7.1 Automobiles

Surface transportation of people and goods may contribute to the modification of the climate in two ways: first, by the emission into the atmosphere of particulate and gaseous exhaust products that can impact on radiation fields and precipitation; and second, by the modification of the reflective and thermal properties of the land surface by highways. Internal-combustion vehicles are a major source of aerosols, such as lead particles, and of gaseous precursors of "photochemical smog" (specifically reactive hydrocarbons and oxides of nitrogen). The absolute and relative amounts of these emissions attributed to motor vehicle exhaust and gasoline and diesel fuels depend upon a number of factors including engine design, condition of maintenance, and mode of driving. By comparison, the emissions from other forms of surface transportation, such as railroads and ships, are negligible.

The extensive expanses of highways and roads in the developed countries may affect the heat budget of the earth in a small way. Depending upon the type of road construction, the surface albedos can be either greater or less than that of the adjacent landscape. A sense of the problem can be ascertained from data for the United States. The total length of U.S. roads is 5.9×10^6 km, consisting of 5.1×10^6 km of rural roads and 0.8×10^6 km of municipal roads (*World Almanac,* 1969). With an average width of 12 meters, the road coverage of the United States is, therefore, about 7.0×10^4 km^2. The area of the United States is 8.7×10^6 km^2. Hence, about 0.8 percent of the United States is covered by roads. Such a change of the land use is not nearly as great in terms of area as the changes due to irrigation and agriculture discussed in Sections 4.4.2 and 7.4, but it is not entirely trivial.

4.7.2
Aircraft

Aircraft are another source of pollution. In 1969 aviation consumed about 6 percent of the world petroleum production (W. B. Beckwith, private communication, 1971; SCEP, 1970, Table 5.9). However, this source is unique in that most of its products are injected at levels in the atmosphere well above the surface. While its insult to ground-based man is thereby reduced, its potential climatic impact can be greater, and somewhat different, than if the same energy were expended near the ground.

Currently, the bulk of the exhaust products from aviation activities are introduced into the high troposphere or, if in the stratosphere, very close to the tropopause. Only the water vapor has been indicated as a possible cause of climate change by the increase of high cloudiness (see Section 8.7.4).

Supersonic commercial aircraft, on the other hand, introduce their exhaust products mainly into the lower stratosphere. The potential effects on climate and on the ozone layer in the stratosphere, which shields life from ultraviolet radiation, are discussed in Chapter 9.

There can be little doubt that the aviation industry will grow in the next 20 to 30 years. Even with an additional fleet of supersonic transport aircraft, the number of passenger-miles carried and the amount of jet fuel consumed by subsonic aircraft will increase about threefold by 1985 (Beckwith, private communication, 1971). However, technological progress may reduce the amount of pollution from each passenger-mile by perhaps a factor of 2. The result may therefore be less than a threefold increase in pollution from aircraft in 1985. Supersonic commercial transports (SSTs) have not yet been placed into operation, but projections, now in some doubt, suggest about 500 SST aircraft in operation by 1985 to 1990.

4.8
The Case of the Arctic Ocean Ice

The pack ice that covers about 90 percent of the Arctic Ocean and extends into the North Atlantic plays a special role in the heat budget of the Northern Hempishere. The reason for its importance is, in brief, that it has a high albedo, so it reflects solar radiation that would, in the absence of the ice, be absorbed by the Arctic Ocean. This permits the ice to perpetuate itself by its cooling effect on the entire Arctic Basin. With this concept in mind, Budyko (1969) and Sellers (1969) have independently suggested that a small warming of the Arctic, by a very few degrees, would start a recession of the ice pack. This melting would cause additional heat to be absorbed, thereby further enhancing the warming trend until the ice had disappeared. A small cooling would, they argue, be amplified in a similar way (see Sections 6.5, 6.7, and 6.8).

This highly simplified explanation of the role of the arctic sea ice is intended merely to call attention here to its importance, and the matter is given the more detailed attention it deserves in Sections 6.5, 7.2, and 7.3. The important point is that it responds very sensitively to any climate change and also tends to reenforce the change. This reenforcement will be referred to later as a "positive feedback mechanism."

The general view of the sensitivity of the arctic ice pack has been held for many years and has inspired a number of suggestions —they can hardly be considered as serious "proposals"—somehow to eliminate the ice pack (for example, Fletcher, 1970). Ostensibly the objective has been to create an open ocean that would warm the lands bordering on it and also create more rainfall in an area that is now quite arid. However, there are reasons to suspect that the overall effects would not turn out to be beneficial. In any case, it seems to us that it would be highly irresponsible to take any such measures until the consequences could be predicted better than is now possible (see the SMIC recommendation on experiments in climate modification in Section 7.3.1).

Probably the most ambitious of these suggestions is the concept of damming the Bering Strait and pumping water from the Arctic Ocean into the Pacific. This would induce an increased flow of warm Atlantic water into the Arctic Ocean, substantially modify the heat budget of this ocean, and significantly modify, or even eliminate, the area covered by sea ice.

Another suggestion has been that of spreading many tons of soot or black dust on the ice by cargo aircraft, thereby increasing the rate of melting during the spring and summer. Since the average thickness of the pack ice is about 3 meters and it now shrinks by about 1 meter during the melting season, the additional heat absorbed by the darker surface might, it has been suggested, complete the process in one or two seasons.

While these are merely suggestions now, they point to the very real possibility that man *could,* if he chose to, modify the global climate deliberately in a very substantial way. We hope that he will make every effort to understand the full implications of his actions before he tries.

References

Bryson, R. A., 1967. Possibilities of major climatic modification and their implications: Northwest India, *American Meteorological Society Bulletin, 48* (3): 136–142.

Budyko, M. I., 1963. *The Heat Budget of the Earth* (Leningrad: Hydrological Publishing House).

Budyko, M. I., 1969. The effect of solar radiation variations on the climate of the earth, *Tellus, 21:* 611–619.

Budyko, M. I., 1971. *Climate and Life* (Leningrad: Hydrological Publishing House).

Budyko, M. I., Drozdov, O. A., and Judin, M. I., 1966. Vliyani hozyaistvenoi deyatelnosti no klimat, Cb, *Sovramenii problemi klimatologii.* Hydrometozdat, L.

Cronin, J. F., 1971. Recent volcanism and the stratosphere, *Science, 172:* 847–849.

Fletcher, J. O., 1970. Polar ice and the global climate machine, *Bulletin of the Atomic Scientists, 26:* 40–47.

Griffin, J. J., Windom, H., and Goldberg, E. D., 1968. The distribution of clay minerals in the world ocean, *Geophysical Scientific Research, 15:* 433–459.

Hannell, F. G., 1969. Research in urban climatology, *Climatological Bulletin, 5:* 51–53.

Knetsch, G. von, Shata, A., Degens, E., Münnich, K. O., Vogel, J. C., and Schazly, M. M., 1963. *Geologische Rundschau, 52:* 587–560.

Lamb, H. H., 1970. Volcanic dust in the atmosphere: with a chronology and an assessment of its meteorological significance, *Philosophical Transactions of the Royal Society, 266:* 425–533.

Lieth, H., 1964. Versuch einer kartographischen Darstellung der Pflanzendecke auf der Erde, *Geographisches Taschenbuch 1964/65* (Wiesbaden: Steiner), pp. 72–80.

Lundholm, B., 1970. Interactions between oceans and terrestrial ecosystems, *Global Effects of Environmental Pollution,* edited by S. F. Singer (Dordrecht, Holland: D. Reidel Publishing Company; New York: Springer-Verlag), pp. 195–202.

Lvovich, M. I., 1969. *Vednie Resoursi Budushevo* (Water Resources of the Future) (Moscow: Prosveschenie), pp. 174.

Lvovich, M. I., 1970. *World Water Balance: General Report,* Symposium on World Water Balance, University of Reading, England, Publication No. 93: 401–415.

Lydolph, P. E., 1970. *Geography of the U.S.S.R.,* second edition (New York: John Wiley & Sons, Inc.), Chapter 15.

Oke, T. R., 1969. Towards a more rational understanding of the urban heat island, *Climatological Bulletin, 5:* 1–20.

Organization for Economic Cooperation and Development (OECD), 1969. Basic statistics of energy (Paris, annually).

Schmidt, W., 1917. Der Einfluss grosser Städte auf das Klima, *Die Naturwissenschaften, 5:* 494–495.

Seddon, B., 1967. Prehistoric climate and agriculture: review of recent paleoecological investigations, *Weather and Agriculture,* edited by J. S. Taylor (Oxford: Pergamon Press), pp. 173–185.

Sellers, W. D., 1969. A global climatic model based on the energy balance of the earth-atmosphere system, *Journal of Applied Meteorology, 8:* 392.

Study of Critical Environmental Problems (SCEP), 1970. *Man's Impact on the Global Environment* (Cambridge, Massachusetts: The M.I.T. Press).

Sulakvelidze, G. K., 1968. Metodi vozdeistviia na gradove protsessy, *Vysokogornyi Geofizicheski Trudy, 11,* 283 pp.

The Times, 1968. *The Times Atlas of the World,* comprehensive edition (London: Times Newspapers Ltd.).

United Nations, 1968. *World Energy Supplies* (New York: United Nations).

U.S. National Academy of Sciences—National Research Council, 1966. *Weather and Climate: Modification Problems and Prospects,* NAS/NRC Publication No. 1350 (Washington, D.C.).

World Almanac, 1969. (New York: Newspaper Enterprise Association, Inc. for Doubleday and Company, Inc.).

Part III The Theory of Climate

5 Factors Governing the Climate

5.1 Introduction

From the description of ways in which man's activities may affect climate, which were enumerated in Chapter 4, it was apparent that such influences are not generally direct ones but arise from the indirect effects of a variety of interacting processes. In order to understand the consequences of changes originating in man's activities, therefore, we shall examine in this chapter the main factors and processes that govern all climatic changes, natural or man-induced. For convenience, these processes will be grouped under the headings of radiative processes, ocean-atmosphere transport processes, and hydrological processes.

The main factors that determine the overall climate of the earth-atmosphere system are the input of solar radiation, the composition of the atmosphere, and the earth's surface characteristics. We shall first discuss the character of solar radiation, the reflective properties of the land or sea surface and various atmospheric constituents, and the absorption and emission of long- and shortwave radiation. It will then be apparent that the motions of the atmosphere and oceans play an essential part in the overall heat balance, transporting both latent and sensible heat from regions of net radiative heating to those of net radiative cooling. Finally, the hydrological factors of climate are briefly reviewed in the light of possible man-induced changes in the water balance.

5.2 Radiative Processes

5.2.1 Incoming Solar Radiation

THE EARTH AND THE SUN

The annual average total incoming solar radiation per unit time at the top of the earth's atmosphere on a unit surface perpendicular to the sun's rays (that is, the solar constant) is 1.95 cal cm^{-2} min^{-1} (0.136 W/cm^2) (Drummond, 1970). Almost all of this radiation originates in the solar photosphere. Although there have been many suggestions of variations of the solar constant (see, for instance, Kondratiev and Nikolsky, 1970) changes of total solar radiation, even of the order of 1 percent, have not been firmly established.

The solar spectrum, especially in the visible and infrared regions, corresponds roughly to the energy distribution of a blackbody of about 5800°K with some irregularities due to absorption in the solar atmosphere. More than 90 percent of the total solar radiation lies in the spectral interval from 4000 Å to 40,000 Å (1 Å is equal to 10^{-8}cm). The maximum solar intensity is in the green portion of the visible spectrum at around 4800 Å. Large variations —corresponding to periods of extreme solar activity—are frequently observed in the x ray and extreme ultraviolet (UV). Recently satellite observations of UV radiation from the sun also indicate intensity variations of 10 to 20 percent in the spectral region from about 1300 Å to 2000 Å, with a period coinciding with that of the solar rotation (Heath, 1969). However, no significant changes of the solar spectrum in the visible or infrared have yet been documented. Similar variations are known to occur in the solar flux at 10.7 centimeters. The microwave part of the solar spectrum, however, carries very little energy, and these latter fluctuations serve only as indicators of solar activity. The atmosphere above about 70 kilometers shows a fairly clear association with these observed variations of solar radiation. No such relationships are evident in the behavior of the stratosphere or troposphere.

The total incoming solar radiation on a horizontal surface at the top of the atmosphere depends also on the earth-sun distance and on the solar elevation angle. The latter two parameters are functions of the time of year and location on the earth and the earth's orbital characteristics, such as the eccentricity of the earth's orbit, the obliquity of the ecliptic, and the longitude of perihelion with respect to the spring equinox. At the present time, the earth's eccentricity is approximately 0.0167, giving a variation of incoming solar radiation of ±3.3 percent about its mean value from perihelion (at the beginning of January) to aphelion (at the beginning of July).

The orbital elements, as calculated from celestial mechanics, show slow variations with average periods of the order of 100,000 years for the eccentricity, 40,000 years for the obliquity, and about 21,000 years for the "effective period of precession" (that is, the longitude of perihelion). The combined variation of these elements produces irregular fluctuations in the total solar radiation received by the earth and at different latitudes. According to Milankovitch (1941), small values for the eccentricity and obliquity, when accompanied by winter solstice near perihelion, are favorable conditions for snow and ice accumulation in polar regions. Although direct verification of climate changes corresponding to these astronomical considerations has not been forthcoming, recent isotope analyses have provided some suggestion that temperature fluctuations during the past 400,000 years may have been influenced by changes in the earth's orbital elements (Emiliani, 1966 a, b). These influences, of course, are completely independent of man's activities.

GLOBAL ALBEDO

A large portion of the incoming solar radiation is reflected back to space without participating in the thermal budget of the earth-atmosphere system. The amount of radiant energy reflected by the earth to space divided by the total incoming solar energy is called the global albedo. This reflection is from clouds, the atmosphere (both molecules and dust particles), and from the earth's surface.

Clouds

The reflectivity of a cloud varies according to its height, its thickness, the size and concentration of the cloud elements, and whether the cloud is composed of ice or liquid water droplets. On the average, high thin cirrus has a reflectivity of up to 20 percent; altostratus and altocumulus of about 50 percent; and cumulus of about 70 percent, although very deep cumulonimbus could have even higher reflectivity. It is possible that large numbers of atmospheric particles in clouds, such as those found over industrial areas, would change these reflectivities somewhat, although this is not yet certain (see Chapter 8).

The average cloudiness over the globe is close to 50 percent (slightly higher in the Southern Hemisphere) and represents the major contribution to the planetary albedo, or reflectivity of the earth (London and Sasamori, 1971; Budyko, 1971). It should be noted that, since the average cloudiness distribution both by type and total amount varies considerably with location and season, its contribution to the earth's local albedo is a very sensitive parameter in the regulation of the climate and of climatic changes. Therefore, the optical characteristics of clouds and a continued census of the cloud amounts by type and height represent fundamental input information to our understanding of the thermal budget of the earth-atmosphere system.

Whether man could change the amount or optical properties of the global cloudiness through his activities is not yet certain, nor is it clear yet how clouds might act as a feedback mechanism to suppress or enhance climatic changes from other sources (see later chapters).

Particles

Although the clouds and the surface are the major factors in determining the global albedo, molecules and particles in the atmosphere do contribute.

A particle both scatters and absorbs incident radiation. If the radius of the particle is small compared to the wavelength of the incident radiation, it behaves in a manner similar to an air molecule; that is, part of the incident radiation is redistributed (scattered) in a symmetric manner. The exact angular distribution is described in this case by what is generally called Rayleigh scattering. The Rayleigh scattering can be calculated with relatively little difficulty and would not be altered by changes in the molecular composition of the atmosphere from man's activities.

As the size of the particle approaches the wavelength of the radiation, the distribution of the scattered radiation becomes increasingly asymmetric and more complicated. The complete theory of the scattering of radiation by a spherical particle was developed by Mie. As the particle size increases, more and more energy is scattered in a lobe around the forward direction.

It is interesting to note that, because of the difference in size and scattering characteristics, molecular scattering produces relatively strong reflectivity in the blue part of the spectrum, whereas larger particle scattering is less dependent on wavelength. (Changes in appearance of the sky from deep blue to milky white indicate an increased amount of dust or haze particles). Mie's theory also allows the computation of the fraction of the incident radiation which is absorbed (transformed into heat) by the particle, provided that the particle size, wavelength, refractive index, and absorption coefficient are all specified.

The ratio of the fraction of radiation scattered by a particle to the sum of that radiation which is both scattered and absorbed, with the condition that light is scattered only once by the particles, is called the single scattering albedo of the particles. Another important parameter is the so-called phase function, which describes the angular distribution of the intensity of the scattered radiation. These parameters depend upon the size distribution of the particles (that is, the relative number of particles per unit volume with radii in a given size range), their refractive index, and their absorption coefficient.

In passing through the atmosphere, radiation is attenuated by gases and particles by both scattering and absorption. The "optical thickness" τ of a layer is an important quantity that describes the opacity of the layer. The unattenuated fraction of radiation passing through a layer is by definition $e^{-\tau}$. Except for very thin layers where τ is small (less than about 0.05), light is probably scattered more than once by the particles, a process called multiple scattering. There are "exact solutions" (that is, numerical solutions to the multiple scattering equations) available to the multiple scattering problem for a particle layer (for example, Hansen, 1969), but these involve considerable numerical computation. In many contexts, simple idealized models are helpful. The simplest, but still useful, model envisages simple upward and downward streams of radiation that can be absorbed and reflected. The fraction of the radiation which is reflected from a particle layer is often called the "backscatter." These approximate solutions of the multiple scattering problem are computed using values of the phase function and single scattering albedo calculated from Mie theory for a layer of particles with given optical parameters (see Rasool and Schneider, 1971).

The ratio of the radiant energy absorbed in the layer to that reflected by the layer, together with the albedo of the underlying surface, determines whether the layer of particles taken together with the underlying surface increases or decreases the global albedo (see, for example, Schneider, 1971).

Although particles contribute at present only a small fraction of the global albedo, their effects are given emphasis here because man's activities are responsible for a sizable fraction of the particles now observed in the atmosphere (see Chapter 8).

Surface

The earth's surface also contributes to the overall albedo of the earth-atmosphere system. The reflectivity is, of course, quite different for different types of surfaces (see also Section 7.4). For example, the reflectivity of water surfaces is approximately 3 to 8 percent and can be more (up to 25 percent in polar regions) depending on the average elevation of the sun and the relative diffusivity of the incident solar radiation affected by, for instance, the amount of cloudiness present. The reflectivity of dark coniferous forests and arable land is 10 to 15 percent. For deciduous forests it varies from about 10 percent during winter to as much as 25 percent but could be as high as 40 percent for extremely dry sandy conditions. Finally, the reflectivity of snow and ice varies from about 30 to 70 percent, depending largely on the age and condition of the surface. In remote polar areas the reflectivity of permanent snow can be as high as 90 percent. Although estimates are available for the total reflectivity of various types of surfaces, its spectral variation is not, in general, well known.

Since most of the unreflected energy is absorbed by the earth's surface, it follows from the preceding discussion that the dark ocean areas absorb much more solar radiation than do the brighter continents. In polar regions there is less energy available at the surface because of the average low elevation of the sun. In addition, much less of the incoming radiant energy is absorbed there because of the high reflectivity of the snow covering. For the same reason much more radiation is absorbed at the earth's surface in the summer hemisphere than in the winter hemisphere—particularly at high latitudes.

The large-scale features of the global albedo are determined by the distribution of land and ocean areas. But to the extent that man can disturb some of the land features (that is, change the soil characteristics or the use for which the land is developed), he can change the energy absorbed by the surface and thus the local input to the thermal budget of the atmosphere (see Chapter 7).

For the entire planet the global albedo of the earth is calculated and observed to be about 30 percent. Clouds contribute about 25 percent to the global albedo, and the atmosphere and the earth's surface contribute the remaining 5 percent (Vonder Haar and Suomi, 1971; London and Sasamori, 1971). The albedo shows distinct variations with cloudiness and with the distribution of continents and oceans.

ABSORPTION OF SOLAR RADIATION

The unreflected part of the incoming solar radiation that does not penetrate to the earth's surface is absorbed in the atmosphere and converted directly into heat (a very small part is used for chemical dissociation and for ionization of the absorbing molecules). In the upper region of the thermosphere, above about 100 kilometers, solar energy in wavelengths below about 1800 Å is screened out largely by molecular oxygen. As mentioned earlier, the mean state of the thermosphere and its temperature variations follow the available energy distribution in the UV solar spectrum and variations of this energy with solar activity. There is no direct evidence, however, that these variations have any influence or association with the dominant physical processes in the lower atmosphere.

Solar radiation at longer wavelengths (that is 1800 Å to 2900 Å) penetrate deeper into the mesosphere and stratosphere, in some cases down to about 20 kilometers. Radiation at these wavelengths is completely absorbed by molecular oxygen and ozone. (As a matter of fact, the ozone peak absorption at wavelengths of about 2550 Å is responsible for the relatively high temperature found at about 50 kilometers). This part of the solar spectrum, however, shows much smaller temporal fluctuations, if any, and associated variations of the temperature in the ozone layer at about 50 kilometers have not yet been observed.

Solar radiation that has not been absorbed in the higher layers (usually wavelengths beyond about 2900 Å) is transmitted to the lower stratosphere and troposphere. This portion of the solar spectrum contains by far (more than 97 percent) the largest portion of the total incoming energy from the sun. On the average only about 18 to 20 percent of the incoming solar energy is absorbed by water vapor, carbon dioxide, molecular oxygen, clouds, and dust in the lower stratosphere and troposphere. More than half of this absorption is due to atmospheric water vapor, mostly in the lower troposphere.

In the lower stratosphere the absorption of solar energy is fairly small, and there is a delicate balance between this absorption and the emitted longwave radiation (discussed in Section 5.2.2). As a result, it may be possible that the existing equilibrium between absorption of solar energy and emission of infrared radiation can be easily disturbed by the addition of new absorbers in that region such as volcanic dust. Since the water vapor distribution in the stratosphere may depend on the temperature of the lower stratosphere (at around 18 to 20 km), the nature and absorption characteristics of particle layers in this region need to be determined (see Chapter 9).

5.2.2
Outgoing Infrared Radiation

THE EARTH AS A RADIATOR

Because it is reasonable to assume that the long-term climatic average of the earth-atmosphere system remains constant, the first step in understanding the nature of climatic changes is to understand the average steady-state climate. For this purpose we consider that the earth-atmosphere system is in radiative equilibrium with the sun. As indicated earlier, only the unreflected part of the incoming solar radiation takes part in the thermal budget of this system. The distribution of absorbed energy varies with height, location, and time. However, the condition of radiative equilibrium for the planet Earth requires only that the total radiation emitted by the earth be equal to that absorbed. The excess or deficit of radiant energy at any particular place or time can be balanced by a change in local temperature, by necessary vertical and horizontal transports by the atmosphere and/or ocean, and by the heat exchange between the atmosphere and its lower boundary. The importance and nature of these dynamical considerations will be discussed in Section 5.3.

An efficiently radiating substance emits radiant energy in

proportion to the fourth power of its absolute temperature. Since the earth is such a substance, it can achieve a state of radiative equilibrium when its temperature is high enough to balance the energy it gains from solar radiation. This temperature is called the effective radiative (or brightness) temperature of the earth. Thus,

$$T_e = \left[\frac{(1 - A)S}{4\sigma}\right]^{1/4}$$

where
A is the planetary albedo (equal to about 0.30), S is the solar constant (1.95 cal $cm^{-2}min^{-1}$), and σ is the Stefan-Boltzmann constant. The factor 4 enters because the solar constant is defined for an area perpendicular to the sun's rays, but the earth as a whole emits radiation through a surface four times as large. This gives a numerical value for the effective radiating temperature of the earth-atmosphere system as $T_e \approx 253°K = -20°C$.

THE ROLE OF THE ATMOSPHERE—THE "GREENHOUSE EFFECT"

The observed mean global temperature at the earth's surface is about 14°C, or approximately 34 degrees warmer than the effective radiative equilibrium temperature. This difference is, of course, due to the effect of the earth's atmosphere and is often called the "greenhouse effect." This effect is analogous to that of the glass of a greenhouse, since the atmosphere permits nearly half the incoming solar radiation to penetrate down to the surface (where it is used to raise the surface temperature or evaporate water), whereas less than 10 percent of the longwave radiation emitted by the surface (which tends to cool the surface) can escape through the atmosphere to space.

The temperature of the earth is very much lower than that of the sun. Therefore, its radiative emission takes place in the longwave or infrared part of the spectrum, between 5 μ and 100 μ, the so-called terrestrial spectrum. In the infrared part of the spectrum, the reflectivity of clouds is quite low (that is, they absorb and emit radiation quite efficiently like a blackbody). Also, water vapor and carbon dioxide are much more opaque to infrared radiation than to visible radiation. As a result, especially in the cloudy areas, radiation from the earth's surface cannot directly escape to space. Instead, radiation from the cloud tops or from the optically active gases is emitted at a temperature much lower than that of the surface. Atmospheric water vapor absorbs the radiation in the spectral interval from 5 to 8 μ and beyond 19 μ; carbon dioxide absorbs in the interval from 12 μ to 18 μ. Spectral regions of weak absorption located between the above-mentioned bands are called the infrared "windows" of the atmosphere—the deepest and largest of these transparent windows extends between 8 to 12 μ.

The atmosphere intercepts the infrared radiation from the surface, and radiates energy both back down to the surface (warming the surface) and upward to space (cooling the planet). In general, therefore, the radiation lost to space comes from the upper layers of the troposphere and not directly from the earth's surface. This so-called "emission layer" is just below the tropopause for water vapor and in the lower stratosphere for carbon dioxide. The temperature at these heights is about 65°C colder than at the earth's surface. A large portion of the outgoing radiation thus comes from cold upper layers of the troposphere, and, compensating for this, the downward radiation from the atmosphere warms the earth's surface above the effective radiative temperature of the planet. Thus, the major part of the outgoing terrestrial radiation originates in the atmosphere and comes from the radiating gases such as H_2O, CO_2, and from clouds and dust.

In the thermal spectrum of the atmosphere a particle layer has little effect in the wavelength regions in which strong selective absorption occurs. Only in the "windows" do the particles have to be considered. Because of the size distribution of typical atmospheric particle layers, the effect of the particles on infrared radiation is usually less pronounced than their effect on solar radiation, but this is not always true (see Section 8.7.2).

It is evident that changes in average cloudiness, carbon dioxide concentration, water vapor, or dust can disturb this balance and thereby effect changes in the climate. Detailed discussion of these effects is left to later chapters, but it is evident that variations in each of these parameters will perturb the so-called "greenhouse" balance. Since the atmosphere is a highly interactive system, all factors will contribute more or less to this perturbation.

ATMOSPHERIC TEMPERATURE PROFILE—THE ROLE OF CONVECTION
Since the surface absorbs nearly half the solar input and is a source of heating, the temperature of the lower atmosphere decreases with height above the surface. The rate of decrease, called the tropospheric lapse rate, is determined by two processes. One is the radiative heating of each atmospheric layer, where the absorption and emission of radiation by optically active gases, clouds, and particles act as a local (*in situ*) source or sink of heat. The other process is vertical convection, which occurs when lower atmospheric layers are heated sufficiently to cause them to expand and become lighter than the surrounding air. Then warm air currents move upward, being replaced by cooler descending streams. This process is also important in the formation of clouds. Convection adjusts the atmospheric temperature profile rapidly, and is most important in the troposphere, whereas radiative heating directly affects the temperature profile primarily in the upper atmosphere. Modification of the temperature profile by injection of particles into the atmosphere is a possibility that should be investigated further (see Section 8.7.2).

5.2.3
Conclusions

The incoming solar radiation at the top of the atmosphere is well known as a function of latitude and time, but there is still some question about its absolute intensity and possible time variations. Also, the earth's albedo and its geographic variations are known generally from satellite observations and model calculations, but accurate values of the albedo and details of its distribution are less well known. The geographic distributions of the incoming solar radiation and albedo determine the absorption of energy by the earth-atmosphere system and thus represent major factors in determining the overall climate. Absorption shows strong variation with geography and season. The highest values are observed over cloudless oceans, lower values over cloudless or partly clouded continents, and the lowest values are found over heavily overcast or snow- and icecovered regions.

Solar energy is absorbed in the atmosphere principally by tropospheric water vapor, and the absorption is much smaller than the infrared emission. Thus, there is a net cooling of the atmosphere insofar as radiative processes are concerned. The amount and vertical distribution of the radiative cooling depends largely on the clouds, water vapor, and the temperature distribution in the atmosphere. The deficit of energy from radiative cooling of the atmosphere is balanced by convection of warm air from the surface and release of latent heat of condensation. The radiative heating at the earth's surface depends on the incoming solar radiation at the surface and the surface reflectivity, and it is generally higher than the outgoing terrestrial radiation (except in the cases of high temperature and surface reflectivity). Thus, for the atmosphere, there is a heat source at the ground and a radiative heat sink within the atmosphere itself. Because of the variations in cloudiness, water vapor, and ground reflectivity, these sources and sinks have large geographic variability.

In general, however, the heat sources are at low elevations in the atmosphere, and the heat sinks are in the upper troposphere. Also, there is an excess of incoming solar radiation over outgoing terrestrial radiation in tropical regions and the reverse in polar regions. Evaporation of water at the ocean surface and condensation in the lower troposphere—particularly in the tropics—provide an additional energy source for the atmosphere. These processes provide a source of available potential energy to drive the atmospheric circulation as a direct heat engine. Since the circulations so developed may themselves be responsible for changes in cloudiness and the transport of heat, water vapor, and dust, this system contains its own feedback mechanisms, and is discussed further in Section 5.3.

5.2.4 Recommendations

1. We recommend monitoring of the temporal and geographical distribution of the earth-atmosphere albedo and outgoing flux over the entire globe, with an accuracy of at least 1 percent.
2. We recommend monitoring with high resolution the global distribution (horizontal and vertical) of cloudiness, and the extent of polar ice and snow cover with somewhat less resolution.
3. We recommend measuring the distribution, optical properties, and trends of atmospheric particles and clouds over the globe.
4. We recommend determination of the absolute value of the

solar constant to better than ±0.5 percent and the spectral distribution of solar radiation from 1800 Å to 40,000 Å to a few percent (±1 percent in the visible part of the spectrum).

5.3 Ocean-Atmosphere Transport Processes

As indicated in the preceding section, the radiative processes in a stationary atmosphere are such that there is a net heating in the tropics and a net cooling at high latitudes. Therefore, the temperature distribution cannot remain stationary unless there is some other mechanism operating within the atmosphere and/or oceans to transport excess heat from the tropics toward the poles. As a second consequence of the radiative imbalance or differential heating, the temperature decreases from the equator to the poles.

In both observation and theory these two consequences of radiative imbalance are intimately connected. It is now well established that strong equatorward gradients of temperature lead to large-scale fluctuating motions of the atmosphere, and that these latter have precisely the effect of transporting some of the excess of heat from the tropics to the polar regions (Thompson, 1961). In the following sections, we discuss this and other important aspects of the large-scale atmospheric and oceanic motions that transport heat, moisture, and momentum.

5.3.1 The Heat Balance

Each column of the earth-atmosphere system has an energy loss or gain (due to the net radiation) that is a function of latitude, longitude, and season. The energy surplus in the column is used to raise the temperature of the air, land, or ocean or to evaporate more moisture from the surface than precipitates in the air column. Any remaining energy may be transported horizontally to other latitudes and longitudes by atmospheric or oceanic currents. Numerous observational studies have given us a knowledge on a zonally averaged basis of the magnitudes of the various quantities involved (see Figures 5.1 and 5.2).

By comparing the net outward flux in the air columns, based on actual measurements of wind and temperatures, with the total amount of energy to be transported out on the basis of heat im-

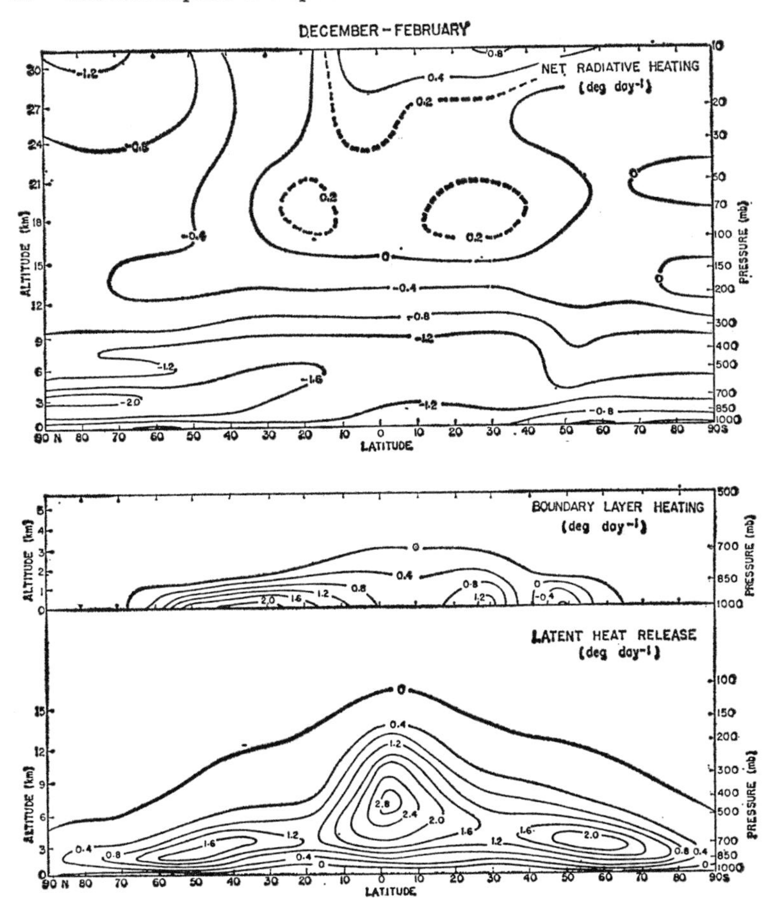

Figure 5.1 Components of diabatic heating for the atmosphere for December to February. Units: degree/day.
Source: Newell et al., 1969.

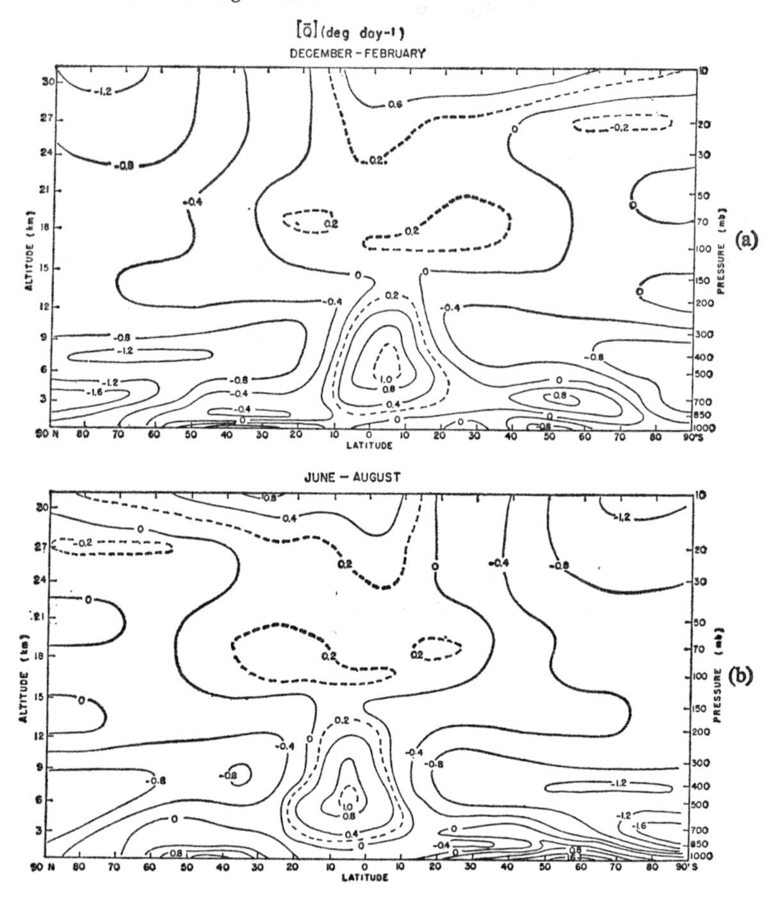

Figures 5.2 a and b Total diabatic heating for the atmosphere for (a) December to February, (b) June to August. Units: degree/day.
Source: Newell et al., 1969.

balance in each of the latitudinal belts, it can be shown that these curves do not coincide (see Figure 5 of Newell et al., 1969). The difference is clearly due to the heat transport effected by the ocean currents, for which there are no direct measurements.

The contribution of the ocean transport to the global heat budget, determined as a residual term, is shown in Table 5.1 (Budyko, 1971). These data indicate that the total mean meridional heat transport produced by the oceans amounts to a few times 10^{22} calories per year in the neighborhood of latitude 30° N.

A single large oceanic fluctuation, however, moving the top 100 meters of water over an area of 10^6 km^2 at a rate of 10 km/day, can transport several percent of this amount in a few months. Such large-scale "turbulent" motions in the ocean seem to be fairly common and can result in temperature perturbations of a degree or so persisting for a year or more. Budyko (1971) has collated the results of several authors, mostly using Atlantic weather-ship data, to show that anomalies in the heat content of the upper layers of

Table 5.1
Zonal Redistribution of Heat by Atmosphere and Ocean (kcal/cm^2/yr)

Latitude Band (°)	Surface Net Radiation In	Evaporation minus Precipitation	Net Atmospheric Transport Out	Net Oceanic Transport Out
70–60° N	−49	−8	−33	−8
60–50	−30	−15	−4	−11
50–40	−12	−9	4	−7
40–30	4	13	0	−9
30–20	14	31	−16	−1
20–10	23	11	2	10
10–0	29	−33	48[a]	24
0–10 S	31	−14	24	21
10–20	28	16	9	3
20–30	20	32	−8	−4
30–40	9	19	−4	−6
40–50	−8	−8	6	−6
50–60	−29	−27	9	−11

Source: Budyko, 1971.
[a] The radiation data do not balance for the latitude band 10° to 0° N. We believe this particular datum should be lowered by 10 kcal/cm^2/yr in order to establish the proper balance.

the ocean amount to about half the magnitude of the annual cycle. In addition, the major meridional currents, such as the Gulf Stream, are known to vary appreciably in strength. Since the heat capacity of the upper 150 meters of the ocean—the approximate depth showing variations over a period of the order of a year—is about 5×10^{22} cal/deg, it is apparent that fluctuations in oceanic transport may be quite a large fraction of the long-term mean and can contribute substantially to the year-to-year variation of climate.

5.3.2
Atmospheric Wind Systems

An early concept of the average motion of the atmosphere was a zonally symmetric motion, with zonally symmetric meridional components whose direction and magnitude depended upon the latitude, altitude, and season. Deviations from such a system of winds were believed to be minor perturbations. However, actual observations and recent theoretical investigations have shown that such a simplified concept of the global wind system is untenable (Lorenz, 1969). Eastward angular momentum is transported from the equatorial latitudes to the middle latitudes by nearly horizontal eddies, 1000 kilometers or more across, moving in the upper part of the atmosphere (8 to 15 km). Such a transport leads to an accumulation of eastward momentum over the middle latitudes, where a strong meandering current of air, generally known as the jet stream, develops. The cores of strong winds in both the hemispheres in both the seasons occur at an altitude of nearly 12 kilometers over the middle latitudes.

More momentum than is necessary to maintain these jet streams against dissipation through internal friction is transported to these zones of strong winds. The excess is transported downwards to maintain eastward flowing surface winds of the middle latitudes against the ground friction. This supply of eastward momentum to the earth's surface tends to speed up the earth's rotation. Counteracting such a continuous speeding up of the earth's rotation, air flows toward the equatorial regions, forming the so-called trade winds. These trade winds, with a strong component directed toward the west, retard the earth's rotation. The air brought to the equatorial latitudes by the trade winds picks up eastward momentum from the earth's surface and transports it to the top of the troposphere.

The trade winds also pick up moisture evaporated from the oceanic surfaces along their trajectories close to the earth's surface. As they ascend over the equatorial latitudes, the water vapor condenses, liberating the heat of condensation to the ascending air. This heating further accelerates the ascent and, in general, all the meridional components of the wind system. The warmed ascending air transports its heat to the higher latitudes principally through the eddies that, as stated earlier, also transport eastward momentum.

In the higher latitudes (poleward from latitudes 30° N and S) nearly horizontal eddies dominate to about 65° N and S, although the transport of angular momentum decreases and finally reverses at about 60° N and S. However, the meridional components (see Figure 5.3) of motion near the surface and in the upper levels are the reverse of those that occur in the lower latitudes. There is a marked poleward component in the winds near the surface and an equatorward component in the upper levels, with an average descending motion in the subtropics and air ascending in the middle and high polar latitudes. The poleward-moving air picks up moisture from the underlying ocean surface and transports it polewards; the condensation of the water vapor in the ascending branch releases latent heat, balancing the deficiencies of heat input in the still higher latitudes. Here again, the precipitation resulting from condensation is an indirect—although for human activity very important—consequence of the poleward transport of heat. Here also the conditions for evaporation and precipitations are highly variable in space and time.

It is thus clear that the large-scale eddies play an essential role in the transport of heat from the equatorial latitudes and give rise to the meandering subtropical jet stream, the trade winds, tropical convection, and copious rain. Rainfall is a secondary effect depending on the *excess heat input* in low latitudes. The latter is used to evaporate water, which is subsequently released as heat of condensation for eventual poleward transport as sensible heat to regions of *deficient heat input.* The rate of evaporation obviously depends upon the ocean surface temperature, which is

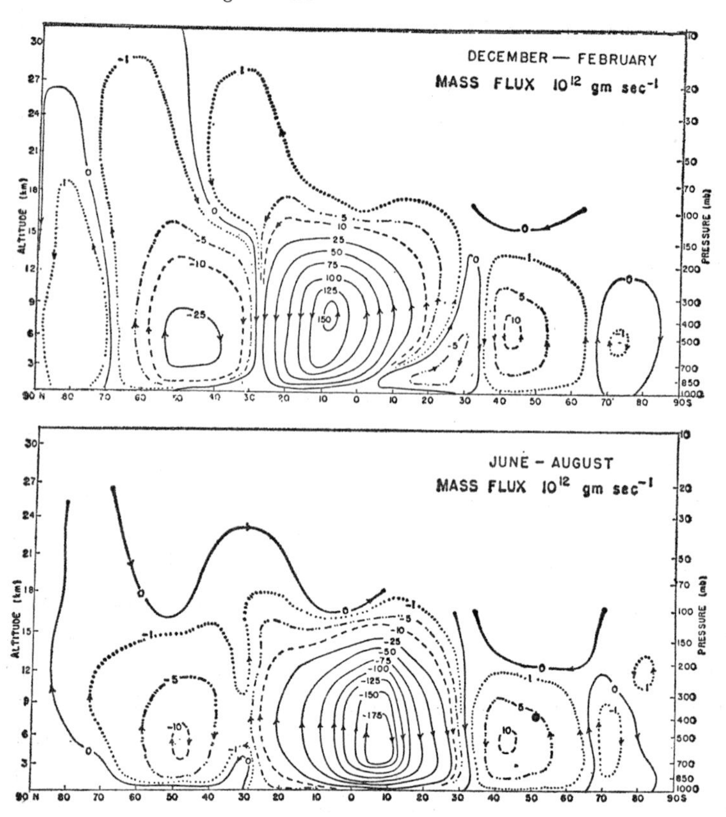

Figure 5.3 Mean meridional circulation for two three-month periods. Units: 10^{12} g/sec
Source: Newell et al., 1969.

partially controlled by the turbulent mixing in the upper layers of the oceans, a process that is still poorly understood. Since the poleward transport of heat in the atmosphere is brought about through nearly horizontal migratory eddies with preferred regions of ascending motion, the rainfall is not uniform in either space or time.

5.3.3 Conclusions

The wind systems prevailing in the atmosphere transport heat from the areas of excess input into areas of deficit input. In this process they generate and maintain the jet streams, the trade winds, the migratory large-scale eddies, the surface eastward winds of the middle latitudes, and the surface westward winds of the polar latitudes. They also account for the clouds and precipitation, which are incidental processes associated with the transport of heat from low latitudes to higher latitudes.

However, the wind systems alone are not capable of transporting the required amount of heat. An amount of heat comparable to that transported by the winds is also transported by ocean currents. The ocean surface temperatures, controlled by turbulent mixing in the upper hundred meters or so of the ocean, play a significant role in determining the exchange of sensible and latent heat between the atmosphere and oceans. Knowledge regarding (1) the role of oceans in the heat budget and (2) the exchange of heat and momentum between the atmosphere and the ocean surface is currently inadequate. Furthermore, information regarding the winds in the equatorial latitudes is also inadequate.

5.3.4 Recommendations

1. We recommend that the surface temperature of the ocean be monitored, preferably on a seasonal basis, if the information is obtainable by remote sensing from satellites.
2. We would like to be able to recommend a monitoring program for the temperature distribution and currents in the upper ocean. However, we recognize that there is at present no economical and effective way to perform such monitoring. Instead, therefore, we recommend combined theoretical and observational stud-

ies to determine the best way to obtain the oceanic data required to compute oceanic transport of heat. (In this connection, efforts should be made to compile climatological data in a single file to aid in the detection of climatic trends in the "state" of the ocean.)
3. We recommend an increase in the number of observations of upper atmospheric winds and temperatures over the tropics.

5.4
Hydrological Processes

Hydrological processes have played an important role in the past history of the earth's climate and are also of basic significance in man's present and potential impact on climate (see also Section 7.4).

It is simplest to regard the hydrosphere as a global system consisting of four reservoirs interrelated by the flux terms of the hydrological cycle (Figure 5.4). Estimates of the amount of water

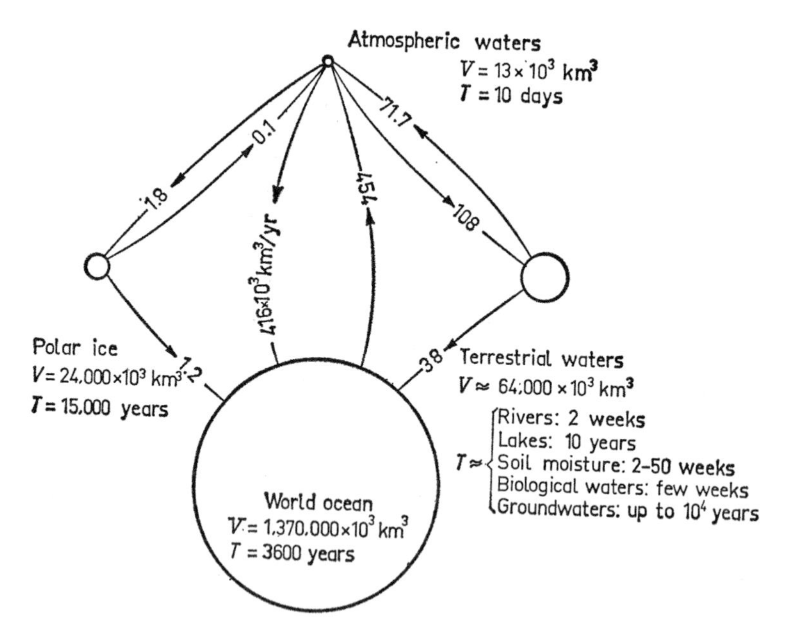

Figure 5.4 Principal statistical and dynamical characteristics of the hydrosphere.
Source: Szesztay, 1971.

Table 5.2
Review of Estimations on Principal Components of the Hydrosphere
(Total water resources of the globe = 1.5×10^9 km^3)

	Area Covered (10^6 km^2)	Present Amount of Water Resources (10^3 km^3)	(percent)	Changes of the Amount during the Last 18,000 Years (10^3 km^3)	80 Years (10^3 km^3)
World ocean	360	1,370,000[a]	93	+40,000[b] (110 m in level)	+100[b] (0.27 m in level)
Polar ice	16	24,000[b]	2		+40[b]
Terrestrial waters	134	64,000[c]	5		
Atmospheric waters	510	13[a]	0.001		

Sources:
[a] UNESCO, 1970.
[b] Orvig, 1970.
[c] Lvovich, 1970, includes groundwaters 64,000; lakes 230; soil moisture 82; rivers 1.2.

Table 5.3
Average Water Balance of the Land Areas (cm/yr)

	Precipitation	Evaporation	Runoff
Africa	69	43	26
Asia	60	31	29
Australia	47	42	5
Europe	64	39	25
North America	66	32	34
South America	163	70	93
Total land areas	73	42	31

Source: Budyko, 1971.

stored in these four reservoirs at the present time are summarized in Table 5.2. The interrelations among the four subsystems or reservoirs can be expressed by the equations of water continuity. The sum of the four storage components is constant since there is negligible creation or destruction of H_2O:

$$\Sigma V = V_{\text{ocean}} + V_{\text{ice}} + V_{\text{groundwater}} + V_{\substack{\text{atmospheric}\\ \text{waters}}} \approx 1.4 \times 10^9 \text{ km}^3$$

Table 5.4
Average Water Balance of the Oceans (cm/yr)

	Precipitation	Runoff	Evaporation	Flow
Atlantic	89	23	124	−12
Indian	117	8	132	−7
Pacific	133	7	132	+8
World ocean	114	12	126	0

Source: Budyko, 1971.

The storage components are connected by the flux terms of water balance: precipitation, evaporation, and runoff. Measurements of the flux terms are presented in Tables 5.3 and 5.4 characterizing the water balance conditions of the major parts of the land and ocean areas.

5.4.1
Variations in Geological Time

During the history of the hydrosphere, the ice ages have resulted in major redistributions of water storage, particularly between the oceans and the polar ice masses. The low ocean level of 18,000 years ago (approximately 110 meters below the present level) was very probably associated with the increase of polar ice masses during the last ice period. There is considerable geological evidence that the significant progressions or regressions of the boundaries of the polar ice caps during the Pleistocene ice ages were accompanied by corresponding fluctuations of the mean ocean level.

It is probable that the subsurface reservoirs of groundwaters were filled by a more or less continuous accumulation process extending through the whole history of the hydrosphere. Climatic changes have affected this accumulation process primarily by shifting the boundaries of the humid zones and changing the rate of deep percolation. Sedimentation in closed inland seas also contributed considerably to the formation of subsurface reservoirs.

The considerable changes in the surface temperatures of the oceanic and land areas have certainly influenced the rate of evaporation and thus the amount of precipitation and the intensity of the transfer mechanisms of the water cycle. The interrelation between temperature regime and turnover rate of the hydrological cycle is one of the areas where research is urgently needed. Moreover, very little is known about the changes in the average amount of atmospheric water vapor content and in the average cloudiness, which probably accompanied variations in turnover rate of the hydrological cycle.

5.4.2
Recent Variations in the Global Hydrological System
Mean sea level has been rising throughout this century. For the period 1890 to 1940, Orvig (1970) calculates a rise of 0.2 meter. Since 1940 the rate of rise appears to have decreased by some 40 percent. It has usually been assumed that the source of this increased oceanic volume—1200 km^3/yr—could be found in the melting of the polar ice caps. However, a recent computation by Orvig (1970) suggests that the polar ice itself is expanding in volume by as much as 500 km^3/yr, the 100-km^3/yr loss in the Arctic being more than compensated by a 600-km^3/yr gain in the Antarctic. The validity of the notion that the ice volume is expanding certainly requires further checking, but in any case it highlights the need for careful examination of the state of nonrecharged groundwater resources. These form the only reservoir large enough to supply any significant demand from the oceans and ice masses.

5.4.3
Conclusions
The intensity of the hydrological cycle may be an important factor in climate change, particularly in view of the evidence of diminishing groundwater supplies.

5.4.4
Recommendation
We recommend an investigation of the interactions between the intensity of the hydrological processes, changes in the storage components of the water balance, and climatic change. Particular emphasis should be given to the examination of these factors in relation to past climate changes.

5.5
General Conclusions

In our present state of knowledge of climatic factors, it is necessary to compare many inferred results with the current state of the atmosphere and oceans and with the distribution of solar and terrestrial radiation in and above the atmosphere. Some of the requirements are for observations of types not previously made by meteorologists and never made on and in the oceans. Some physical properties of the atmosphere, particularly those of clouds and the dust load, are also not known in sufficient detail. We therefore find it necessary to recommend certain data-gathering programs and physical investigations to support the study of climatic theory and the development of models. Our recommendations for research, observation, and monitoring have been summarized at the end of each section of this chapter.

5.6
Some Implications for Implementation of the Recommendations

Satellite techniques appear to be mandatory for implementation of Recommendations 1 and 2 of Section 5.2.4 and Recommendation 1 of Section 5.3.4, and they appear to be the most economical method of making the continuing observations called for in Recommendation 4 of Section 5.2.4. It is our impression that the sensing and other techniques required to make these observations with the necessary accuracy (long-term stability of standardization is called for) are available now. At least one satelliteborne experiment is in preparation now which will test in operation equipment that could satisfy a portion of the requirements of these recommendations, with the possible exception of Recommendation 2 of Section 5.2.4. A continuous service of operational meteorological satellites has now been maintained for some years and these vehicles are of a degree of sophistication that could accommodate the appropriate sensors. Using these satellites would, of course, require reconsideration of some details of the role planned for them, and perhaps some readjustment of priorities within the meteorological community. The intellectual and material expenditure that these recommendations require is not out of proportion in relation to the total now devoted to academic and operational meteorology.

Recommendation 2 of Section 5.3.4 is of a different type. It refers to the "state of the oceans," and calls for a study rather than an observation program only because of our failure to specify a feasible program, which in turn stems from ignorance of the nature of the motions and structure of the oceans on a wide range of scales. We are agreed that ultimate development of detailed theories of climate will require a considerable extension of observational oceanography. We are not at present able to formulate even an estimate of minimum requirements.

The recommendation on particles is amplified in the discussion of this problem in Chapter 8.

References

Budyko, M. I., 1971. *Climate and Life* (Leningrad: Hydrological Publishing House).

Drummond, A. J., 1970. Precision radiometry and its significance in atmospheric and space physics, *Advances in Geophysics, 14:* 1–52, edited by H. E. Landsberg and J. Van Mieghem (New York: Academic Press).

Emiliani, C. E., 1966a. Paleotemperature analysis of Caribbean cores P-6304–8 and P-6304–9 and a generalized temperature curve for the past 425,000 years, *Journal of Geology, 74:* 109–125.

Emiliani, C. E., 1966b. Isotopic Paleotemperatures, *Science, 154:* 851–857.

Hansen, J. E., 1969. Exact and approximate solutions for multiple scattering by cloudy and hazy planetary atmospheres, *Journal of the Atmospheric Sciences, 26:* 478–487.

Heath, Donald, F., 1969. Observations of the intensity and variability of the new ultraviolet solar flux from the Nimbus III satellite, *Journal of the Atmospheric Sciences, 26:* 1157–1160.

Kondratiev, K. Y., and Nikolsky, G. A., 1970. Solar radiation and solar activity, *Quarterly Journal of the Royal Meteorological Society, 96:* 509–522.

London, J., and Sasamori, T., 1971. Radiative energy budget of the atmosphere, *Man's Impact on the Climate,* edited by W. H. Matthews, W. W. Kellogg, and G. D. Robinson (Cambridge, Massachusetts: The M.I.T. Press), pp. 141–155.

Lorenz, E. N., 1969. The nature of the global circulation of the atmosphere: a present view, *The Global Circulation of the Atmosphere,* edited by G. A. Corby (London: Royal Meteorological Society), pp. 3–23.

Lvovich, M. I., 1970. *World Water Balance: General Report,* Symposium on World Water Balance, University of Reading, England, Publication No. 93: 401–415.

Milankovitch, M., 1941. Kanon der Erdbestrahlung und seine Anwendung auf das Eiszeitproblem (Canon of Insolation and the Ice Age Problem), Royal Serbian Academy (Jerusalem: Program for Scientific Translations).

Newell, R. E., Vincent, D. G., Dopplick, T. G., Ferruzza, D., and Kidson, J. W., 1969. The energy balance of the global atmosphere, *The Global Circulation of the Atmosphere,* edited by G. A. Corby (London: Royal Meteorological Society), pp. 42–90.

Orvig, S., 1970. *The Hydrological Cycle of Greenland and Antarctica,* Symposium on World Water Balance, University of Reading, England, Publication No. 92: 41–49.

Rasool, S. I., and Schneider, S. H., 1971. Atmospheric carbon dioxide and aerosols: effects of large increases on global climate, *Science, 173:* 138–141.

Schneider, S. H., 1971. A comment on climate: the influence of aerosols, *Journal of Applied Meteorology, 10:* 840–841.

Szesztay, K., 1971. The hydrosphere and the climatic changes, to be printed in No. 2. Natural Resources Forum, United Nations, New York.

Thompson, P. D., 1961. *Numerical Weather Analysis and Prediction* (New York: The Macmillan Company), p. 118.

United Nations Educational, Scientific and Cultural Organization, 1970. *World Water Balance: General Scientific Framework of Study,* technical paper No. 6 by T. G. Chapman, S. Dumitrescu, R. L. Nace, and A. A. Sokolov (Paris: UNESCO).

Vonder Haar, T., and Suomi, V. E., 1971. Measurements of the earth's radiation budget from satellites during a five-year period, *Journal of the Atmospheric Sciences, 28:* 305.

The General Theory of Climate and Climate Modeling

6.1
The Physical and Mathematical Basis of Climate Theory
6.1.1
Qualitative Discussion

The processes discussed in Chapter 5 determine the changing condition of the atmosphere, oceans, and underlying solid surface in considerable detail and, consequently, determine changes in their average condition. Thus, a quantitative and complete mathematical description of the way in which these processes operate constitutes a general theory of climate. Our next concern is to outline a systematic approach to the theory of climate through the physical principles that describe each process or mechanism.

Viewing the atmosphere first as an isolated system with prescribed input of solar radiation and hypothetically known surface effects, the most relevant principles are the following:

1. Newton's second law of motion, relating the acceleration of any mass to the forces acting on it.
2. The principle of conservation of mass.
3. The first law of thermodynamics, relating the heat added to any gas to the change in its temperature and the work done by it in expanding against pressure forces.
4. The laws of radiative transfer, expressing the rates of heat absorption and emission to the molecular structure and concentration of the various atmospheric constituents.
5. The principles of diffusion and thermodynamics of water vapor.

All of these, it will be noted, describe processes of change. It should also be pointed out that these principles can be expressed in mathematical form by a set of coupled partial differential equations (see Section 6.1.2). Under the hypothetical conditions we have temporarily assumed, the number of equations in that set is exactly equal to the number of variables that appear in the set (with one reservation that will be discussed shortly) Another property that is pertinent from the standpoint of solving the equations is that each equation of the set expresses the change of one variable of the set in terms of the current values of only the variables that appear in the set. Thus, with prescribed solar radiation and surface effects, the detailed evolution and, *a fortiori,* changes of the average state of the atmosphere are given by the solution of the equations just described.

This statement is true in principle but requires a reservation in practice. In actuality, it is not possible, nor will it ever be possible, to solve the fundamental equations in such detail that one can treat explicitly the effects of very small scale motions. The latter are known to be of crucial importance in the turbulent transport of heat, momentum, and atmospheric constituents; they are, however, quasi-random and their statistical effect is related to average conditions over periods of the order of a day. In practice, therefore, the equations discussed earlier are reformulated to apply to variables that are averaged over periods that are large compared with the time scale of rapidly fluctuating motions but small compared with the time scale of large features of the atmosphere's circulation.

The problem of relating the statistical effects of very small scale motions to conditions on a much larger scale raises the more general question of relating the net effect of all microphysical processes to quantities that are more easily observed and computed. This is the so-called "parameterization" problem, one that will be discussed in more specific terms later in this chapter.

As indicated in our qualitative discussion of the main factors of climate change in Chapter 3, the picture of an isolated atmosphere is highly overdrawn, simply because the effects of the oceans and underlying solid surface are changing and depend on variable conditions of both surface and atmosphere. Without going into elaborate detail, however, we merely point out that the condition of the oceans and solid surfaces is governed by physical principles analogous (and in many respects identical) to those that control the behavior of the atmosphere. These are all susceptible to mathematical expression and lead to a more complete system of equations in which the number of equations is equal to the number of variables. These describe not only the state of the atmosphere but also the state of the oceans and underlying soil or snow surface.

The totality of physical and mathematical principles that describe the evolution of the atmosphere and the underlying surface comprises the basis for a theory of climate. Before discussing various ways in which this basis might be applied to the study of climate, we shall discuss in Section 6.2 the possibility that more than one climate may correspond to a single set of conditions external to the atmosphere.

6.1.2
Summary of Equations Governing Atmospheric General Circulation

This statement of the governing equations of a planetary atmosphere will appeal perhaps to the mathematically inclined with some knowledge of physics, because it is a concise formulation of a highly complex problem. The same statement would require many pages of words. Some of the relationships are explicit, and some are left as unspecified functions of a set of independent variables because the formulation of the complete expression is uncertain—these are relationships that could be written "in principle." The first set can be expressed explicitly, and they are the "equation of motion,"

$$\frac{\partial \mathbf{V}^*}{\partial t} + \mathbf{V}^* \cdot \nabla \mathbf{V}^* + w^* \frac{\partial \mathbf{V}^*}{\partial z} + \left[\frac{1}{\bar{\rho}} \frac{\partial}{\partial z} \overline{(\rho w' \mathbf{V}')}\right] + \mathbf{K} \times f \mathbf{V}^* + \frac{1}{\bar{\rho}} \nabla \bar{p} = 0 \tag{6.1}$$

the "equation of continuity,"

$$\frac{\partial \bar{\rho}}{\partial t} + \nabla \cdot \bar{\rho} \mathbf{V}^* + \frac{\partial}{\partial z} (\bar{\rho} w^*) = 0 \tag{6.2}$$

the "hydrostatic equation,"

$$\frac{\partial \bar{p}}{\partial z} + g \bar{\rho} = 0 \tag{6.3}$$

the "equation of state,"

$$\bar{p} = R \bar{\rho} T^* \tag{6.4}$$

the "definition of potential temperature θ,"

$$\theta^* = T^* \left(\frac{p_0}{\bar{p}}\right)^K \tag{6.5}$$

and the "first law of thermodynamics,"

$$\frac{\partial \theta^*}{\partial t} + \mathbf{V}^* \cdot \nabla \theta^* + w^* \frac{\partial \theta^*}{\partial z} + \left[\frac{1}{\bar{\rho}} \frac{\partial}{\partial z} \overline{(\rho w' \theta')}\right] = \left[\frac{1}{\bar{\rho}} \overline{\rho H}\right] \tag{6.6}$$

where $\mathbf{V}$ is the horizontal velocity vector; w is the vertical velocity; p, T, and ρ have their usual connotations; g is the acceleration of gravity; f is the Coriolis parameter, $2\,\Omega \sin \varphi$, Ω being the angular velocity of the earth and φ being latitude; $\mathbf{K}$ is a unit vector in the vertical or z direction; and K is the ratio of the gas constant for air to the specific heat of air at constant p, R/C_p. The superscript * means a density-weighted average, while the conventional $^-$ means an average over a period of time small compared to the characteristic time of large-scale synoptic processes but large compared to that of small-scale turbulent motions. The heat H added to a volume of the atmosphere is given by

$$H = \text{Rad.} + \text{Cond.} \tag{6.7}$$

which is the combined effects of radiational heating and the release of latent heat of condensation. These are functionally related to the other variables and the boundary conditions imposed by the sun and the earth's surface as follows:

$$\text{Rad.} = \text{Rad.}\,(\bar{p}, T^*, q_i^*, \text{solar rad., surface}) \tag{6.8}$$

$$\text{Cond.} = \text{Cond.}\,(\bar{p}, T^*, q_w^*, w^*) \tag{6.9}$$

where the relative concentration, or mixing ratio, of the ith optically important minor constituent q_i (q_w is for water vapor) can be determined by the equation of continuity for that substance:

$$\frac{\partial q_i^*}{\partial t} + \mathbf{V}^* \cdot \nabla q_i^* + w^* \frac{\partial q_i^*}{\partial z} + \left[\frac{1}{\bar{\rho}} \frac{\partial}{\partial z} \overline{(\rho w' q_i')}\right] = \frac{1}{\bar{\rho}} \overline{\rho S_i} \tag{6.10}$$

in which the source (sink) for the ith substance

$$S_i = S_i(\bar{p}, T^*, \mathbf{V}^*, w^*, q_i^*) \tag{6.11}$$

Now, if we set the expressions in square brackets equal to zero, we would be left with seven relationships (Equations 6.1 through

6.6, and note that Equation 6.1 will in fact be two equations because there are two components of **V**), and seven unknowns (two components of $\mathbf{V}^*, w^*, \overline{\rho}, \overline{p}, T^*, \theta^*$).

This provides a closed set of equations that could describe the dynamics of a simplified dry model atmosphere—but certainly not the climate. Such a model has been used to good advantage to make numerical forecasts.

Going a step further, consider what happens when the term on the right in Equation 6.6 involving H is put back, which will in turn require introducing q_i^* and the corresponding Equations 6.10 and 6.11 that govern them, a pair of equations for each q_i^*. The set of equations can *still* be considered as closed, though more difficult to solve.

The more advanced numerical models that are being developed in the United States and the Soviet Union are incorporating a few of these additional relationships in order to obtain more realism, and the vertical eddy transport terms (in square brackets) are introduced by various forms of parameterization. The problems revolving around the treatment of some of these factors are discussed in the text.

6.2 Determinism of Climate

6.2.1 Discussion

It is generally assumed (1) that the system consisting of the atmosphere as a whole and its external environment determines the climate in a more or less unique fashion, (2) that the physical laws governing the system remain unchanged, and (3) that the most likely cause of a change in the long-term behavior of the atmosphere is due to a change in its environment. This last point will be examined here.

It is reasonable to assume that today's gross climate would continue to prevail as long as the present external environment that is not influenced by the atmosphere is unchanged. By including in the same system the global atmosphere and all the parts of its changing immediate environment, then the expected climate changes would be purely internal changes, and in a single model experiment over a long enough time climate changes would appear.

The instantaneous physical state of the atmosphere-ocean-earth system is characterized by a finite set of dependent variables, the so-called variables of state considered as functions of the space coordinates and time. Mathematically, the physical laws governing the system are expressed as a closed system of nonlinear differential equations, each of which expresses the time derivative of one physical entity in terms of the instantaneous state of the system (see Section 6.1.2). Such a system has a form suitable for stepwise numerical integration. The nonlinearity of the differential equations arises from the transport processes (advection) of atmospheric and oceanic properties by the motions of air and water. Having carried out numerical integration of the fundamental equations over a long interval of time, we can then form long-term statistics of the numerical solutions in order to determine the climate as defined by the long-term statistical properties of the state of the atmosphere.

Strictly speaking, long-term statistics are statistics taken over an *infinite* interval of time. On the other hand, the time-dependent solution of a closed system of differential equations defining the behavior of a physical system is uniquely determined by an arbitrarily specified set of initial conditions. Such a solution extends infinitely in time so that it determines a set of long-term statistics. Thus, when defining a climate, if the initial conditions are slightly but randomly modified while the governing equations remain the same and there is a nonzero probability that the resulting climate will be unchanged, then the climate is said to be *stable.* If the initial conditions are similarly modified but there is a zero probability that the resulting climate will be unchanged, then the climate is said to be *unstable.* A system possessing only one single stable climate, and possibly many unstable climates, is called *transitive.* A system possessing more than one stable climate is said to be *intransitive.* The occurrence of a particular climate of an intransitive system is most likely fortuitous.

Until now we have identified the climate with a set of statistics extending over an inifinite time interval. This mathematical concept of climate is not in keeping with the usual notion of climate as defined by a set of statistics extending over a long but finite time interval. Indeed, a climate defined over an infinite interval of time does not change.

The system of differential equations describing the behavior of the atmosphere-ocean-earth system probably has periodic solutions of widely different time scales and it is conceivable that a particular solution extending over an infinite interval of time has quite different sets of statistics over long finite intervals. Such a system is strictly transitive, but in Lorenz's (1968) terminology it is *almost intransitive,* since "statistics taken over infinitely long time intervals are independent of initial conditions, but statistics taken over very long but finite intervals depend very much upon initial conditions." Examples of almost intransitive systems may be found in the literature (Lorenz, 1965; Kraus and Lorenz, 1966), but it has not yet been demonstrated that the real atmosphere is almost intransitive. It is conceivable, however, that almost intransitivity could be a "cause" of climate change. This should be verified using simplified models simulating a hypothetical earth with constant environment, for example, an earth with uniform surface devoid of oceans, ice, and dust. In this way it might be established that long-term climate changes may result from almost intransitivity of the atmosphere rather than from changes external to the atmosphere, so that the mere existence of long-term climate changes could not be taken as proof of environmental changes (Lorenz, 1968).

Owing to the present uncertainty about the uniqueness and transitivity of stable climatic states, our current estimates and judgments regarding man's impact on climate are equally in some doubt. Accordingly, efforts should be made to establish quantitatively the degree of intransivity of climatic states.

6.2.2
Recommendation

We recommend that realistic atmosphere-ocean-earth model experiments be carried out to demonstrate:

1. The uniqueness or nonuniqueness of stable states of climate under identical external conditions.

2. The possible existence of long-period climate changes not caused by changes external to the atmosphere.

6.3 Theoretical Models of Climate

With the reservations implicit in the preceding section, let us now turn to alternative formulations of climatic theory, varying as to physical content and mathematical complexity. To summarize them briefly, the distinguishing characteristics of these alternatives are the following:

1. "Global-average" models, in which the variables that describe the climatic state are independent of location but apply to average values taken over the entire atmosphere or the entire earth.
2. "Statistical" models, in which the climatic state is characterized by variables that have been averaged with respect to longitude but may still be functions of latitude and time. In such models the effects of the atmosphere's large-scale motions are not treated in detail but are treated in a statistical fashion.
3. Semiempirical models of climate based in part on empirical relations derived from direct observation.
4. Generalized numerical models of climate, in which the effect of the atmosphere's large-scale motions is treated explicitly and which may include the effects of all important physical processes in parameterized form.

These types of models will be discussed in the next four sections. The importance of "feedback mechanisms," which can act to suppress or accentuate climate change, are discussed, where applicable, in relation to these models.

6.4 Global-Average Mathematical Models

The simplest kind of mathematical model of the climate considers only globally averaged values of the physical variables. For example, although the solar radiation incident at a given place on the earth is actually a function of the time of year and the location on the globe, in these models the sun's radiation is assumed to be distributed uniformly over the entire globe. Thus the redistribution of absorbed solar energy over the planet by winds, clouds, and ocean currents is not included in such a model. Furthermore, the earth-atmosphere system, because of the large heat storage capacity of the oceans, does not respond instantaneously to changes in the radiation balance that might be induced by a pollutant. The response time of the climate to a small change in the radiation balance is about 10 to 100 years (Manabe and Bryan, 1969). Nevertheless, global-average models are extremely useful in simulating certain aspects of climate and man's impact on climate, especially when a pollutant affects primarily the radiation balance of the earth-atmosphere system.

6.4.1 Changes in Albedo

To determine the influence of the incoming solar radiation and albedo on the mean temperature of the earth's surface it is necessary to know the dependence of the outgoing longwave radiation on temperature. This dependence may be established in two different ways. The first requires the synthesis of theoretical models of the vertical distribution of radiation fluxes, temperature, and composition of the atmosphere.

A second method is based on empirical analysis of data on outgoing radiation derived from observations. The latter method was used in the studies of Budyko (1969), which were the results of calculations of mean monthly values of the outgoing radiation obtained from maps of the *Atlas of the Heat Budget of the Earth* (Budyko 1963). The resulting dependence can be expressed by the empirical formula

$$I_s = a + bT - (a_1 + b_1T)n$$

where I_s is the outgoing radiation, T is the surface temperature, n is the cloud cover, and a, b, a_1, and b_1 are empirical coefficients.

To complete this simple formulation, we shall use also the condition of radiation balance for the globe:

$$Q_s(1 - \alpha_s) = I_s$$

where Q_s is the solar radiation at the outer extremity of the atmosphere and α_s is the global albedo.

With these simple formulas, one can arrive at some general conclusions about influences on the mean temperature near the earth's surface. A change of solar radiation of 1 percent, with cloudiness equal to 0.50 and present global albedo, changes the mean temperature of the earth by 1.5°C. This value is approximately twice as large as the corresponding influence in the absence of an atmosphere, indicating that the radiative properties of the atmosphere considerably increase the potential influence of a change in solar radiation on temperature near the earth's surface (see Section 5.2). A change in the global albedo of 0.01 changes the mean temperature by 2.3°C. Thus, the simplest global average model can show that surface temperature depends very strongly on the albedo variations (provided such variations are not strongly connected with cloudiness).

6.4.2
Changes in Composition

The effect of changes in the composition of the atmosphere on the radiation balance can be computed using a model in which incoming solar radiation, planetary albedo, surface temperature, vertical distribution of temperature, and vertical distribution of optically active atmospheric constituents all have values that are averaged over the earth's surface.

The utility of these models can be illustrated with specific examples. To determine the effect of increased CO_2 concentration on the surface temperature, it is first necessary to demonstrate that changes in CO_2 will affect the radiation balance of the earth-atmosphere system. Then, if there are no other simultaneous or coupled adjustments in other heat balance variables (such as water vapor content in the atmosphere, cloudiness or snow cover) due to the change in radiation balance that might be caused by increased CO_2, the effect of CO_2 changes on the global-average equilibrium atmospheric temperature could be predicted by this model.

Manabe and Wetherald (1967) have constructed a global-average model of radiative-convective equilibrium, in which earth-average values for the vertical distribution of optically active gases are taken into consideration. The global-average atmospheric temperature profile is computed from the model under the following assumptions: (1) the amount of incoming solar radiation absorbed by the earth-atmosphere system is balanced by an equal amount of longwave infrared radiation emitted to space: (2) the amospheric temperature gradient does not exceed the neutral gradient of 6.5°K/km for convection. As discussed further in Section 8.8.2, increases in CO_2 are found to affect the radiation balance sufficiently to suggest the need for more detailed modeling, including important feedback mechanisms, so that the climatic effects of changes in the radiation balance arising from increased CO_2 can be predicted more accurately.

In fact, it is possible to include major feedback mechanisms in a global-average model. For example, if the relative humidity of the atmosphere remains nearly constant while the atmospheric temperature increases (Möller, 1963), then more water vapor will be present in the atmosphere. Since water vapor then acts in concert with CO_2 in increasing the "greenhouse effect," the coupled action of increased CO_2, increased atmospheric temperature, and increased atmospheric water vapor constitutes a positive feedback mechanism, which has already been included in the model by Manabe and Wetherald (1967).

The effect of aerosols on the radiation balance can also be estimated by global-average models. Rasool and Schneider (1971) have found that the effect of aerosols on the global radiation balance can be significant. Similar results are obtained by Yamamoto and Tanaka (1971). Although the ultimate effect of increased aerosols on climate (discussed in Section 8.7) depends in a complex way on many coupled processes in the earth-atmosphere system, these global-average models clearly demonstrate that the magnitude of this effect could be substantial, and that further and more detailed modeling and measurements are necessary.

6.4.3
Conclusion

The use of global-average models, which require very little computer time in comparison with large-scale dynamical models, are a necessary first step in determining how a specific pollutant, possibly introduced by man, might have an impact on climate.

6.4.4
Recommendation

We recommend that global-average models be developed to provide a rough estimate of the effects of a pollutant on the radiation balance. Possible feedback mechanisms and coupled effects that can be modeled on a global-average scale should be included insofar as possible to improve these estimates. When a global-average model indicates that the effect of a particular pollutant on the radiation balance is significant on a global scale, more extensive multidimensional dynamical models should be developed to simulate the impact on climate.

6.5 Feedback Mechanisms

In order to examine the sensitivity of the climate to a given amount of a pollutant, it is first necessary to determine the immediate effect of the pollutant on the radiation balance. It is then necessary to establish how other coupled processes or "feedback mechanisms" might act to suppress ("negative" feedback) or to accentuate ("positive" feedback) that effect on the climate.

To differentiate natural climate variations from those due to man's activity, it is essential to take into account any feedback mechanisms that could be connected with the stability of climate.

6.5.1 Radiative Temperature Change

One of these feedback mechanisms, radiative temperature change, was mentioned earlier during discussion of the global-average models; that is, if the energy absorbed in the earth-atmosphere system is increased, then the mean temperature also increases. But increased temperature is likewise accompanied by increased emission of infrared energy to space, reestablishing the radiative balance between incoming and outgoing energy—but at a higher temperature. The increased emission of radiation from the higher temperature can be thought of as a negative feedback mechanism, where the increased infrared emission acts to limit the temperature increase that would result from additional incoming energy.

6.5.2 Water-Vapor–"Greenhouse" Coupling

In the absence of an atmosphere, the influence of the incoming solar radiation on the temperature near the earth's surface is determined by the Stefan-Boltzmann law. According to this law, an increase of incoming radiation of 1 percent corresponds to an increase of mean temperature of the earth's surface of approximately 0.7°C. In the presence of an atmosphere, an increase of temperature usually causes an increase of absolute humidity (Möller, 1963), which, in turn, considerably increases the influence of radiation changes on surface temperature. This can be identified as a positive feedback case. This case is discussed in more detail in Sections 6.4.2 and 6.8.2.

6.5.3
Polar Ice

A second important feedback mechanism stems from the interrelation between the temperature of the atmosphere and the extent of polar ice. C. E. P. Brooks (1950) suggested that, because of the large albedo of the polar ice caps, they could considerably decrease the temperature of the air above the surface. In studies by Budyko (1961, 1969, 1971) and others it was shown that the increase of absorbed radiation after the melting of some of the polar ice could lead to a potential instability. Calculations made in these studies showed that a small change of heat input to the earth's surface might cause very large variations of polar ice area and strong variations of temperature at middle and high latitudes.

Variations of this kind might arise from the large difference between the albedo of the polar ice and the surface of the ocean. In Figure 6.1 values of incident radiation in the central Arctic, Q_s, radiation absorbed in the earth-atmosphere system, $Q_s\ (1 - \alpha_s)$, and the radiation balance of the earth-atmosphere system, R_s, are presented. It is easy to see from this figure that the radiation excess of the earth-atmosphere system in the Arctic is negative during all the year including summer months, when the incoming radiation is very large. This is explained by the very large value of the albedo of the earth-atmosphere system in regions with permanent polar ice. The negative value of the radiation excess is compensated by the flux of heat from lower latitudes: by the horizontal transfer of heat in the hydrosphere, oceans, and the atmosphere, including redistribution of the latent heat of vaporization (see Section 5.3). Calculation of the arctic heat balance components show that the heat input due to sea currents is a small part of the radiation balance. The difference between heat input of condensation and expenditure of heat of vaporization in the cen-

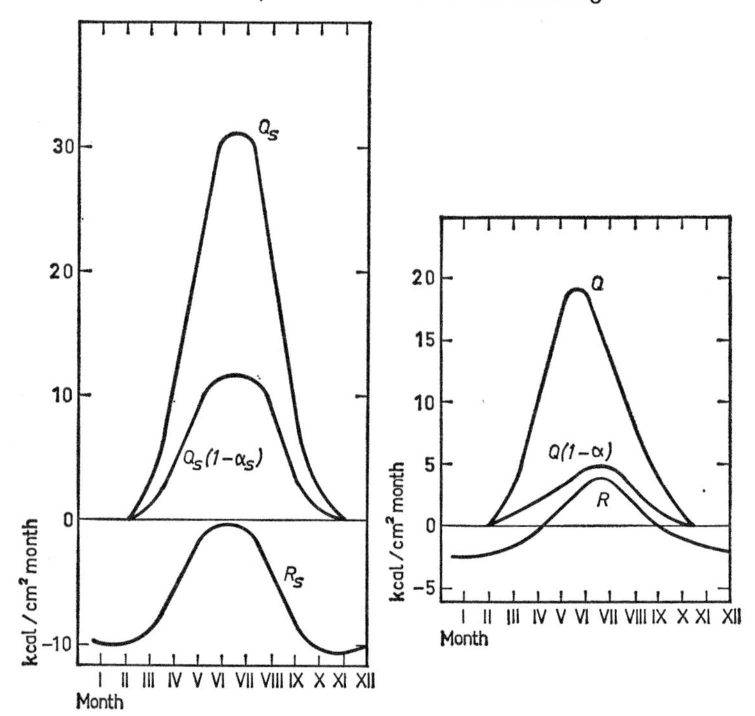

Figure 6.1 Radiation regime of the earth-atmosphere system in the Arctic. Here Q_S represents values of incident radiation in the Central Arctic, and R_S represents the radiation balance of the earth-atmosphere system. The radiation absorbed in the earth-atmosphere system is represented by the expression $(Q_S(1-\alpha_S))$.
Source: Budyko, 1971.

Figure 6.2 Radiation regime of the surface of the sea ice in the Arctic. Here Q represents values of incoming radiation and R represents radiation by the earth's surface. The radiation absorbed is represented by the expression $Q(1-\alpha)$.
Source: Budyko, 1971.

tral Arctic is also comparatively small, no more than a few kcal/cm^2/yr. Thus, the negative radiation excess of the earth-atmosphere system in the Arctic is compensated for mainly by a meridional flux of sensible heat in the atmosphere, which is especially large in the winter months.

In Figure 6.2 values of incoming radiation Q, radiation absorbed $Q(1 - \alpha)$, and radiation by the earth's surface R are presented. This figure makes it clear that in the summer months the ice surface receives a considerable amount of incoming solar radiation because of the long polar days. At the same time solar radiation absorbed by the ice surface and by the earth-atmosphere system in the Arctic is relatively small due to the high albedo of snow and ice. Observations show that in the central Arctic the albedo in spring and autumn is approximately 0.8, decreasing in summer at the time of snow melt to 0.7.

If the polar ice were removed, the albedo of the oceanic surface replacing it would be considerably smaller and the absorbed radiation increased. In a number of independent studies (Budyko, 1961; Rakipova, 1966; Fletcher, 1966; Donn and Shaw, 1966) the possibility of restoring the ice cover after its removal has been discussed in some detail (see also Section 7.3.1). Computation of the heat-balance components has shown that the increase of the air temperature in the Arctic after removing the ice might be enough to prevent re-formation of the ice cover, that is, even for conditions that are near to the present-day climate the ice cover might be unstable. This instability could be a controlling factor during any process of change in the climate.

It is not difficult to include the feedback mechanism connected with the dependence between humidity of the air and a temperature in a theory of climate. It is much more difficult, however, to take into account a feedback mechanism connected with the interrelation of polar ice and the temperature of the air. This has been done only recently using a greatly simplified theory of climate (see Section 6.7.2).

6.5.4 Global Cloudiness

It has been emphasized that global-average models are useful for identifying the first-order effects of changes in a given parameter on the radiation balance. Another application of global-average models is considered next in discussing the possible effects of changes in cloudiness on the radiation balance. The possibility of the clouds acting as a feedback mechanism is outlined, and the need for the use of more detailed models is clearly illustrated by this example.

Approximately one-half of the earth is covered with clouds (London and Sasamori, 1971). The clouds play a major role in the radiation balance of the earth-atmosphere system, since they emit infrared radiation to space in proportion to the cloud-top temperature, absorb infrared radiation originating below their level, and because they reflect a significant fraction of the incoming solar radiation; cloud albedo is about 50 percent when averaged over all types of clouds for the entire earth (London and Sasamori, 1971; Budyko, 1971). Since a 1 percent change in solar energy absorbed by the earth can change the surface temperature by about 1.5°C (see Section 6.4.1), it is clearly important to determine how cloudiness might be affected by changes in the radiation balance resulting from pollution.

Changes in cloudiness, either by a variation in the amount of cloud cover and/or cloud height, is a possible feedback mechanism that should be included in climate models. Since low clouds and middle-level clouds are black to infrared radiation (London, 1957) the cloud tops effectively replace the underlying surface as a black infrared radiator. But since the clouds tops are considerably colder than the surface and atmosphere below, they radiate less energy to space than would be emitted if the clouds were absent (see Section 5.2 for a general discussion of atmospheric radiative transfer).

The importance of clouds in the radiation balance can be demonstrated by using a global-average model. Using the model of Rasool and Schneider (1971), F_{IR}, the infrared flux to space emitted by the earth-atmosphere system for a particular atmospheric temperature profile, can be computed as a function of cloud cover for several different values of cloud height. Figure 6.3 (Schneider, private communication) shows this relation. For each particular cloud height the infrared flux to space decreases with increased cloud cover, since as cloud cover increases a larger

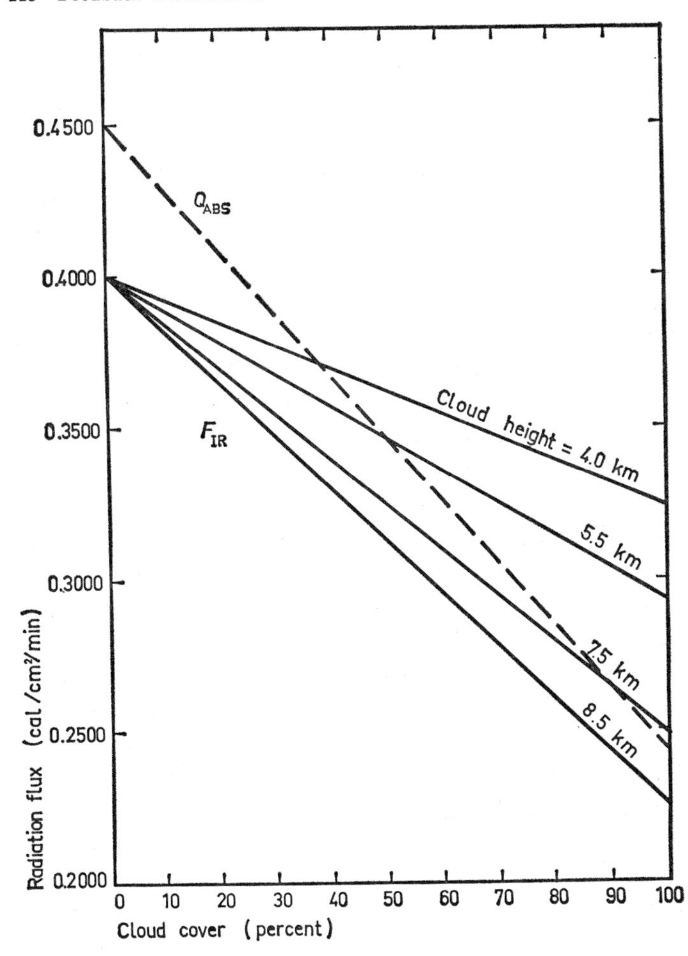

Figure 6.3 Infrared flux to space emitted by the earth-atmosphere system (F_{IR}) for a fixed atmospheric temperature profile, and absorbed solar energy (Q_{ABS}) as a function of cloud cover, for several values of cloud top height. Source: Schneider, 1971, private communication.

fraction of the earth's surface (held fixed at 288°K in the model) is covered by clouds whose tops radiate infrared radiation like a blackbody at the temperature of the atmosphere (about 6.5°K/km colder than the surface) at the cloud-top level. Furthermore, for a fixed amount of cloud cover, F_{IR} will decrease with increasing cloud-top height, since the temperature of the troposphere decreases with height (see Section 5.2.2). Thus, the clouds have a significant effect on infrared radiation.

Clouds, however, are also excellent reflectors of incoming solar radiation. Thus, if the cloud amount increases, not only will the infrared flux to space decrease (which would tend to increase the surface temperature) but the absorbed solar energy decreases because increased cloud cover tends to increase the global albedo (thus decreasing the earth's temperature).

A model for studying which of these competing influences has more influence on surface temperatures can be based on global-average conditions. For an assumed global-average surface albedo (assumed in the model to be 10 percent) and a global-average cloud albedo (50 percent in the model), the earth's albedo can be computed as a function of cloud cover. Then, Q_{ABS}[Q_{ABS} = solar constant × (1 − global albedo)], the amount of solar energy absorbed in the earth's atmosphere system (based on a global average which is one-fourth of the normally incident solar radiation), can also be plotted in Figure 6.3 as a function of cloud cover. Since the cloud albedo is much greater than the average surface albedo, Q_{ABS} also decreases with cloud cover. For radiative equilibrium, Q_{ABS} must exactly equal F_{IR}, otherwise the surface temperature would change. That is, when $Q_{ABS} > F_{IR}$, the temperature increases, or if $F_{IR} > Q_{ABS}$, the temperature decreases until radiation equilibrium is restored.

The intersection of curves Q_{ABS} and F_{IR} determines a set of different possible equilibrium values of cloud heights and cloud cover amounts consistent with constant (in this case, 288°K) surface temperature. For the global-average conditions of 50 percent cloudiness, the average or effective cloud height is found from Figure 6.3 to be 5.5 km. If the clouds were to rise and spread in such a way that F_{IR} always equaled Q_{ABS}, then a constant surface temperature could be maintained. However, if the cloud amount is increased but the height remains unchanged at 5.5 km, the infrared flux to space (F_{IR} for 5.5 km) will then exceed the absorbed solar energy (Q_{ABS}) at the larger cloud cover values, and the atmospheric temperature would be forced to drop in order to restore radiative equilibrium ($F_{IR} = Q_{ABS}$). Similarly, if the cloud cover remained at 50 percent but the height increased above 5.5 km, then Q_{ABS} would exceed F_{IR} and the temperature must increase.

The conclusion that increasing the cloud cover decreases the absorbed solar energy more than it reduces the infrared flux to space is supported by the global-average model of Manabe and Wetherald (1967) for the cases of low- and middle-level clouds, which have high albedo.

Of course, the magnitude of the temperature response due to changes in cloudiness depends on the numerical values of cloud albedo and on the degree of "blackness" of the cloud to the infrared radiation from below. For example, for thin cirrus clouds, the albedo is relatively low (about 20 percent according to Haurwitz, 1948), and these clouds may be only "half black" to infrared radiation. In this case the temperature response to increasing cirrus cloudiness is negligible (Manabe, 1970). In addition, if the cirrus clouds are fully black, then the surface temperature could even be increased by increased cloud cover, provided that the more abundant and brighter low- and middle-level clouds would remain unchanged while the amount of fully black cirrus clouds increases (see also Sections 8.7.6 and 8.9).

Global-average models, however, are probably not sufficient to determine even the direction in which the surface temperature will change with changes in cloudiness. For example, in order to study the influence of clouds on the mean global temperature, Budyko (1971) has compiled empirical data on the distribution over the globe of cloud types, heights, and albedos. His conclusion is that the dependence of surface temperature variations in cloudiness depends critically upon the amount of incoming solar radiation. That is, at high latitudes, where the average incoming solar radiation is below the global-average value used in the models described earlier, clouds tend to have more influence on the infrared radiation than on the absorbed solar radiation, thus increasing the surface temperature. However, for lower latitudes, where the incoming radiation is greater than the global average, clouds usually decrease the surface temperature. Budyko concludes that for the mean global condition the effect of clouds on the solar and infrared radiation fluxes is approximately compensated, and that the influence of clouds as a feedback mechanism is probably less significant than the effect of water vapor or polar ice.

6.5.5
Conclusions

Changes in temperature from changes in heating also cause a change in the same direction of infrared emission. This leads to damped temperature variations from changes in heating and is a negative feedback mechanism.

On the other hand, increased atmospheric temperature could lead to increased water vapor in the atmosphere, thus enhancing the surface temperature change. Therefore, the water-vapor–"greenhouse" coupling is a positive feedback mechanism.

The high albedo of ice cover is an important positive feedback mechanism. For example, colder temperatures would increase the ice cover, which increases the albedo, which further decreases the temperature.

It is possible that clouds could act as a feedback mechanism in response to changes induced by the addition of pollutants to the atmosphere (for example, CO_2-induced temperature rise), but the direction of feedback remains to be determined. Models using realistic distributions of clouds (location over the globe, average cloud-top height at that location and albedo) must be developed to perform the simulation with sufficient accuracy.

6.5.6
Recommendations

1. We recommend that the monitoring programs called for in Recommendation 2 of Section 5.2.4 be implemented.
2. We recommend that simplified climate models be developed to study possible feedback mechanisms, as discussed in Sections 6.4.4, 6.7.2, 6.8.2, and 6.8.6.
3. We recommend that cloud data be incorporated into more detailed climate models (taking into consideration at least the

latitudinal variation of incoming solar radiation and surface temperatures) to study how change in cloudiness might affect the radiation balance, both locally and averaged over the globe.

6.6 Statistical Formulations of Climate Theory

Another class of climate models that are only slightly more complicated than those discussed in the preceding section (and which have the advantage of requiring little computing time) is based on statistical formulations of the fundamental dynamical equations. By averaging variables with respect to longitude and assuming that the most important nonlinear interactions are those between the mean and fluctuating motions, one may formulate the averaged equations as a complete system of equations in which the variables are statistics of the variables that describe the detailed state of the atmosphere. These new variables are average temperature, average wind velocity, net transport of heat, momentum transport, variance of temperature and eddy kinetic energy—all defined at several discrete levels in the atmosphere and considered as functions of latitude and time.

Several models of this type have been designed and tested by Kurihara and Holloway (1967), Saltzman (1968), and Blinova (1943), and the statistical transport properties of the large-scale motions have been investigated by Green (1970). Although they do not apply directly to the question of climate change, the results of these studies are very promising and show that even the simplest models of this type are capable of reproducing the most important features of the general circulation of the atmosphere.

The deficiencies of the statistical models constructed so far are that the surface conditions were assumed fixed and arbitrarily prescribed, and that no attempt was made to discriminate between the relatively short response time of the atmosphere and the very long response times of ocean and underlying solid surface. Recently, however, Thompson (personal communication) has suggested an extremely simple statistical model, in which it is assumed that the atmosphere adjusts instantaneously to changing surface conditions. If such assumptions are found to be valid, they would enormously simplify the calculation of climate changes over very long periods of time, unobscured by the much more rapid changes of the responding atmosphere.

The further development of statistical models appears to be an important aspect of research in the theory of climate from several points of view. First, they are simple enough that it is possible to predict the gross effects of varying a number of parameters and to establish the sensivity of the model's total behavior to small variations of each parameter. This capability would provide invaluable insight in the design of more complex models. Finally, the same kind of insight is needed in interpreting the rather staggering output of nonstatistical models. Models of the latter type, in which the effect of the large-scale motions is treated explicitly, will be discussed in Section 6.8.

6.7
Semiempirical Theories of Climate

6.7.1
Discussion

To study variations of the climate per se it is necessary to develop a model that takes into account all essential components of the heat balance of the earth-atmosphere system and all important feedback mechanisms. This can be done in simple fashion by using empirical relations that permit the construction of parameterized climate models. We shall next consider how such a simplified theory can be used to gain a general understanding of the causes of climate variations (Budyko, 1969). Let us start with the equation of heat balance of the earth-atmosphere system in the form

$$Q_s(1 - \alpha_s) - I_s = C + B_s$$

where C is the rate of heat redistribution due to horizontal movements in the atmosphere and hydrosphere, and B_s characterizes the rate of gain or loss of heat in the system. For mean annual conditions B_s is zero and C is the radiation excess of the earth-atmosphere system. Values of C can be determined from observational data. Let us start by supposing that the values of C are in some way connected with the horizontal distribution of air temperature. To study this relation, let us compare the mean latitudinal value of the annual radiation excess of the Northern

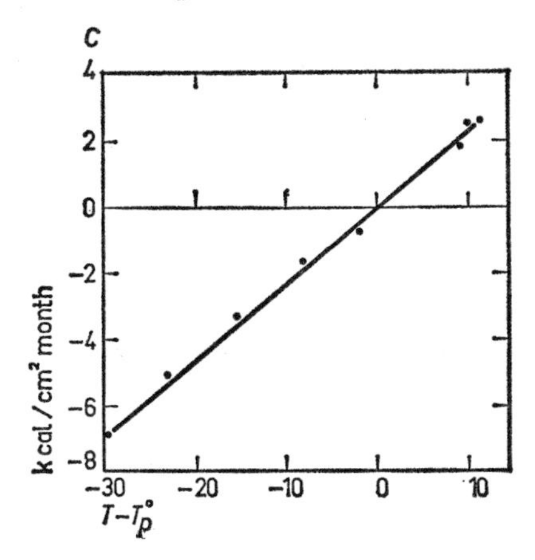

Figure 6.4 Dependence of meridional redistribution of heat in the earth-atmosphere system on differences of temperature.
Source: Budyko, 1971.

Hemisphere with $T - T_p$, where T_p is the mean (global-average) air temperature near the earth's surface and T is the mean air temperature near the surface for different latitude zones. Relations between these values are presented in Figure 6.4. It is seen from this figure that there is a definite relation between C and $T - T_p$.

From this relation, together with the equation of the heat balance of the earth-atmosphere system, it is possible to calculate mean annual temperatures at different latitudes. The distribution of this temperature agrees surprisingly well with the distribution found by observation.

6.7.2
Instability of Polar Ice

To determine the influence of polar ice on climate (see also Sections 6.5.3, 7.2, and 7.3), let us suppose that the border of the polar ice corresponds to a definite air temperature. With this assumption, and using the relationships in Section 6.7.1, we can conclude that the influence of radiation variations on the thermal regime is considerably increased by changes in the area of the polar ice cover. If, for constant albedo, a decrease of radiation of 1 percent reduces the mean temperature of the globe by 1.5°C, then, taking account of the increase of albedo due to increase of the polar ice

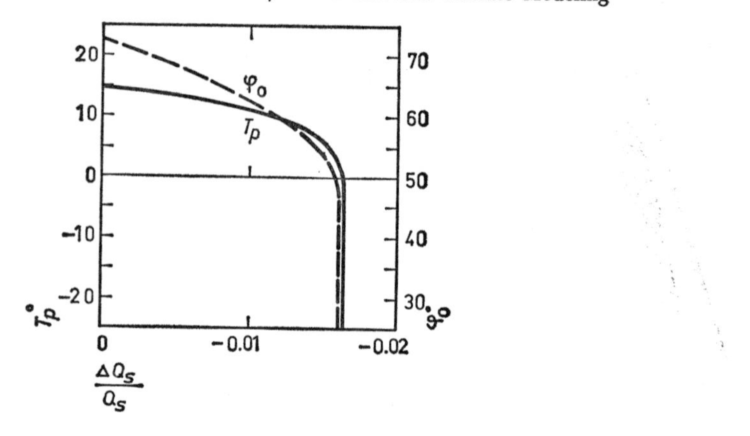

Figure 6.5 Dependence of the mean border of ice cover φ_0 and mean global temperature T_p on the relative change of radiation ($\Delta Q_s/Q_s$).
Source: Budyko, 1971.

area, we find that a decrease of radiation by 1 percent leads to a 5°C decrease. Simultaneously the border of the polar ice would shift to the south by 8 to 18° of latitude, that is, by a distance approximately corresponding to the quaternary glaciations. With a decrease of radiation of 1.6 percent, the ice cover would approach a mean latitude of 50°C, after which it would shift to lower latitudes and reach the equator as a result of self-development, without a further decrease in solar radiation. This change would bring about a sharp decrease of mean global temperature, which would become approximately −70°C. Thus the polar ice exhibits the characteristics of a positive feedback mechanism (Section 6.5.3). The values of mean global temperature T_p at a latitude corresponding to the border of glaciation and its dependence on incoming radiation are presented in Figure 6.5.

It should be mentioned that the notion of very strong instability of the polar ice has also been suggested by Sellers (1969), who also used a semiempirical model of climate based on the equation of the heat balance of the earth-atmosphere system.

More recently, a simplified model of climate has been used to study quaternary glaciation (Budyko and Vasischeva, 1971). For this purpose, it was necessary to calculate the distribution of temperature for different seasons. In this case the value of B_s in the equation of heat balance for the earth-atmosphere system must be taken into account. To calculate this component, the equation of the heat balance of the ocean surface was used.

6.7.3
Factors Affecting Polar Ice

This model described at the end of Section 6.7.2 was applied to a calculation of the thermal regime and polar ice position, including the changes in the earth's orbit elements and the earth's axis inclination which have taken place over the past few hundred thousand years. Some results of the calculations are presented in Table 6.1.

It is clear from the data given in Table 6.1. that the variations of radiation regime caused by changes in the earth's surface position with respect to the sun may lead to substantial climatic changes.

According to this simple model the mean latitudinal limits of ice cover in the glacial epochs shifted 8° to 13° of latitude to the south from the present boundary, located at 72° N. In the Southern Hemisphere in those epochs the limit of ice cover, now situated at 63° N, shifted to the north by a smaller amount, not exceeding 5°. Moreover, according to the calculations described

Table 6.1
Climatic Change in Glaciation Epochs

Time in Thousands of Years before 1800 A.D.	Δ° N	Δ° S	$\Delta T°_{65° N}$
22.1	−8	−5	−5.2
71.9	−10	−3	−5.9
116.1	−11	−2	−6.5
187.5	−11	0	−6.4
232.4	−12	−4	−7.1

Note:
Δ° N = displacement of the mean latitudinal limit of ice cover in the Northern Hemisphere
Δ° S = displacement of the mean latitudinal limit of ice cover in the Southern Hemisphere
$\Delta T°_{65° N}$ = temperature change at 65° N
Source: Budyko and Vasischeva, 1971.

earlier, the periods of maximum glaciation in the Northern and Southern Hemispheres did not coincide. It will also be noted that the maximum calculated value for the extension of the ice cover in the Northern Hemisphere agrees well with observation.

As an example, let us now consider the possibility of a substantial climatic change as a result of comparatively small initial changes in energy input or in air temperature. The calculations carried out using the present model (Budyko, 1971) point to the conclusion that small changes of heat input to the earth's surface or of the mean planetary air temperature may be sufficient to lead to considerable shifts of the limits of polar ice. Thus, with a few tenths of a percent increase in solar energy input or with a few tenths of a degree C rise in the mean air temperature at the earth's surface, north polar ice may completely disappear and south polar ice may recede by 3° of latitude. With a 0.25 percent increase in radiation or with a sustained 0.4°C rise in temperature, this model suggests that the south polar ice might completely melt. These estimates apply strictly to stationary conditions and imply that the computed ice-cover changes can be achieved only by long-term anomalies in the values of energy input or air temperature. It is evident that to change the position of large continental glaciers, which possess great thermal inertia, a very long period of time is needed, but that sea ice (the mean thickness of which does not exceed several meters) can change its position markedly in time periods not longer than a few decades. In particular, the area of north polar ice was somewhat reduced in the twenties and thirties of our century when the mean air temperature at the earth's surface rose by several tenths of a degree.

According to these arguments, a change of mean planetary temperature of the order of 0.5°C or a change in energy input by several tenths of a percent could lead to climatic changes of the same magnitude as those that took place in the twenties and thirties. However, unlike the warming in the first half of our century, which stopped in the forties, a sustained warming might result in the melting of sea ice in the Arctic and might be accompanied by climatic changes over a considerable portion of our planet.

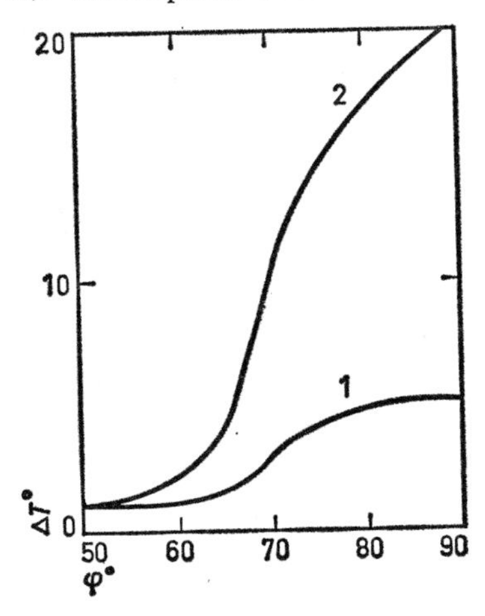

Figure 6.6 Change of the mean temperature at different latitudes for different seasons. Line 1 represents the warm half-year and line 2 represents the cold half-year for temperate and high latitudes in the Northern Hemisphere. Source: Budyko and Vasischeva, 1971.

The character of these changes is shown in Figure 6.6 in which there are shown theoretically computed changes of mean latitudinal temperature in the warm half-year (line 1) and the cold half-year (line 2) for temperate and high latitudes in the Northern Hemisphere. With the disappearance of north polar ice, the air temperature in high latitudes might increase by as much as 20°C in the cold period of year and up to 5°C in the warm one. Warming in latitudes lower than 50° probably would not exceed 1° both in the warm and cold period of the year.

6.7.4
Conclusion

According to a highly simplified semiempirical theory of climate, it has been shown that in the next hundred years the climate of the earth might conceivably change as a result of man's activities. This change may be of importance for continued human activity, especially at temperate and high latitudes of the Northern Hemisphere.

6.8
More General Mathematical Models—Explicit Treatment of Large-Scale Dynamics

6.8.1
Atmospheric Circulation Models

MODEL STRUCTURE

Following the pioneering work of Phillips (1956) and Smagorinsky (1963), intensive efforts have been devoted to the simulation of climate by mathematical models of the atmosphere (see, for example, Mintz, 1965; Leith, 1965; Smagorinsky, Manabe, and Holloway, 1965; Manabe, Smagorinsky, and Strickler, 1965; Kasahara and Washington, 1967). The remarkable development of electronic computers has been one of the major factors responsible for the rapid advance in the field of the numerical simulation of climate. In this section, we shall describe very briefly the structure of a typical model of the climate that capitalizes on the extraordinary power of modern computers.

In a numerical model, the instantaneous state of the atmosphere is represented by the three-dimensional fields of temperature, pressure, water vapor, and wind velocity. The continuous fields of these variables are replaced by their values at a prechosen three-dimensional grid of points. Partial derivatives in the governing equations are replaced by finite differences.

In general, a mathematical model consists of five major parts: the prognostic equations of motion, of temperature (thermodynamical equation), of water vapor, and the balance equations of heat and of water at the ground surface. The equations of motion are usually written in a spherical coordinate system or other conformal map coordinate systems. Since the horizontal scale of the atmospheric motions with which we are concerned is far larger than the scale height of the atmosphere, the hydrostatic approximation is used. The effects of mountains are incorporated as a lower boundary condition.

The scheme of computing radiative heating (or cooling) consists of two parts: absorption of solar radiation and emission of terrestrial infrared radiation. The distributions of radiative fluxes are computed as functions of the distribution of water vapor, carbon dioxide, ozone, and clouds, which are the major absorbers in the atmosphere. The temperatures of continental surfaces are determined in a way that satisfies the requirement of heat balance at the surface. In most of the climate models constructed so far, observed distributions of sea surface temperature are prescribed as a lower boundary condition.

The prognostic system for water vapor consists of contributions by the three-dimensional advection of water vapor, condensation, and evaporation from the earth's surface. The hydrology of the ground surface is taken into account by a prognostic scheme that includes the effects of snow cover and soil moisture.

Figure 6.7 shows the couplings between the five major components of the model described earlier. For example, the equation of motion is related to the thermodynamical equation through the equation of state. The distribution of the wind field obtained from the time integration of the equation of motion is used for the computation of the advection terms in the thermodynamical equation and the prognostic water vapor equation. The heat of condensation computed from the prognostic equation of water vapor constitutes a heat source term in the thermodynamical equation. The computation of radiative transfer yields the rate of radiative heat-

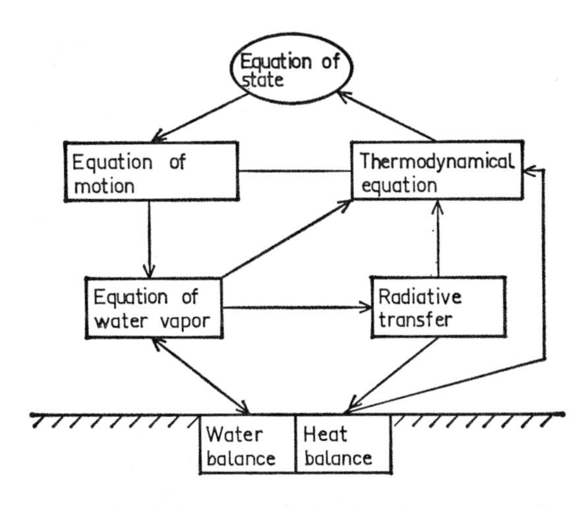

Figure 6.7 Block diagram indicating the structure of a typical mathematical model.

ing (or cooling), which also constitutes a source (or sink) term in the thermodynamical equation. The distribution of water vapor required for the computation of radiative transfer can be obtained from the prognostic equation of water vapor. For the determination of the temperature of the ground surface, the radiative fluxes evaluated in the computation of radiative transfer are used. The computation of ground hydrology requires information on the rates of precipitation and evaporation. The former can be obtained from the prognostic equation for water vapor and the latter from the heat balance computation for the ground surface.

The numerical time integration of a model is usually started from a very simple initial condition, such as an isothermal stationary atmosphere. A model "climate" is expected to emerge from the time integration after a period of 200 to 300 model days, which is a typical time period for the adjustment of the thermal structure of a model atmosphere.

SIMULATION OF CLIMATE

In order to illustrate the performance of numerical models of the atmosphere, we shall show very briefly how numerical models can simulate the global climate. One of the early attempts to simulate the distribution of climate was made by Mintz and Arakawa (Mintz, 1965). Although their model did not incorporate explicitly the processes of the hydrologic cycle, they were able to simulate some of the fundamental features of climate. In Figure 6.8, the distribution of surface pressure resulting from the time integration of their model is compared with the observed distribution of surface pressure in January (note that, in the model, the January distributions of sea surface temperature and solar radiation are used). This comparison indicates that the major features of the distribution of surface pressure in middle and high latitudes are reproduced by the model. For example, the Siberian high, Icelandic low, and low-pressure zones in the North Pacific Ocean are simulated very successfully.

It is well known that the hydrologic cycle is one of the fundamental factors that control the climate (see Sections 5.4 and 7.4), and many investigators have attempted to simulate the hydrologic process in numerical models of the atmosphere (Leith, 1965; Manabe, Smagorinsky, and Strickler, 1965; Manabe, 1969a; Washing-

Figure 6.8 Upper half: the computed time mean distribution of surface pressure. Lower half: observed monthly mean distribution of surface pressure in January.
Source: Manabe, 1968.

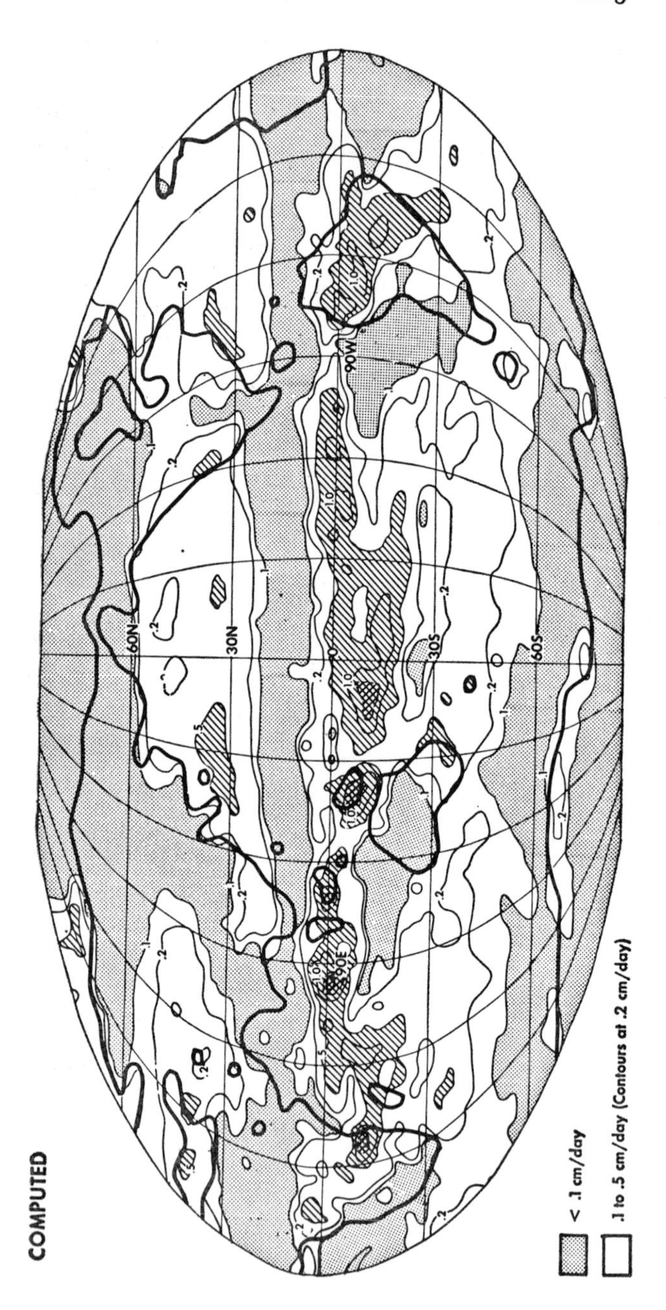
COMPUTED
60N
30N
30S
60S
90W
90E
< .1 cm/day
.1 to .5 cm/day (Contours at .2 cm/day)

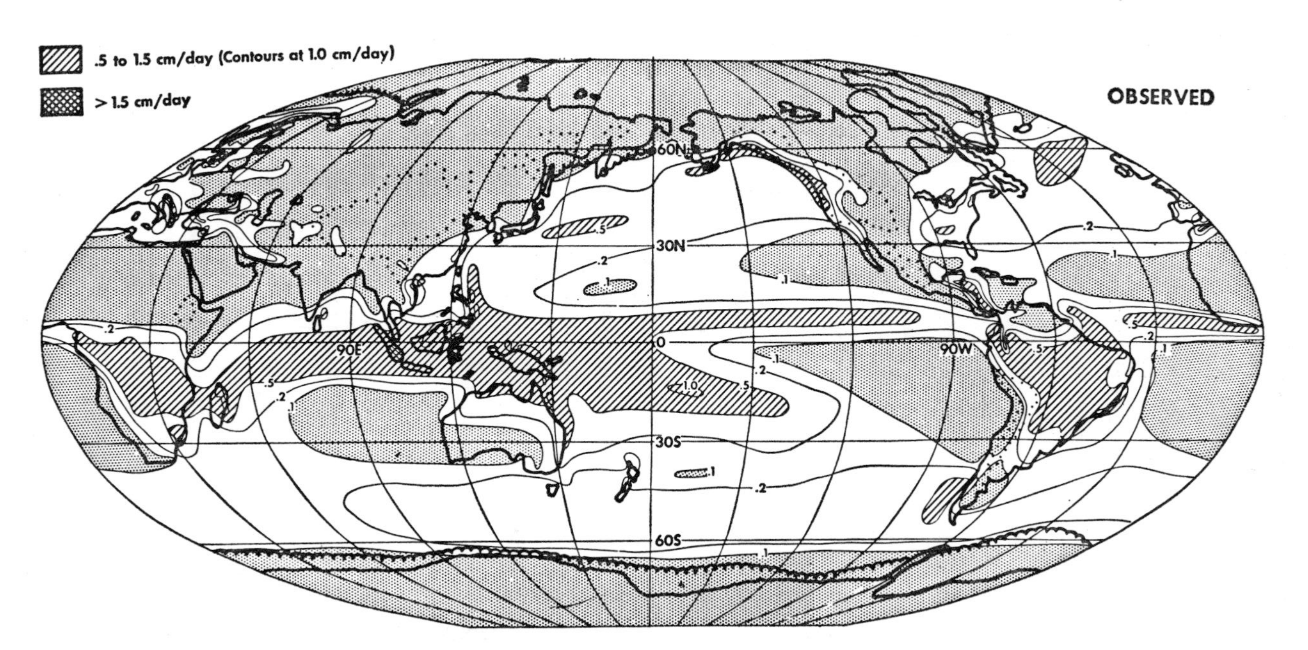

Figure 6.9 Mean precipitation rate computed by the model of Holloway and Manabe compared with the estimated observed rate. Source: Holloway and Manabe, 1971, for the "computed" section; Möller, 1951, for the "observed" section.

ton and Kasahara, 1970; Holloway and Manabe, 1971). In Figure 6.9 the horizontal distribution of the rate of precipitation obtained from one numerical model of the climate (Holloway and Manabe, 1971) is compared with the distribution of the actual January precipitation rate estimated by Möller (1951). (Note that the time integration of the model was performed for January conditions.) This comparison indicates that the model is capable of simulating major arid zones, such as the Sahara and Australia, as well as general features of the distribution of tropical rainfall.

6.8.2
Models for the Study of Climatic Change

So far we have described mathematical models of climate in very general terms. In order to use mathematical models for the study of climatic change, however, it is necessary to incorporate into the models certain interactive mechanisms that determine the sensitivity of climate. Since these interactions are often missing from the existing numerical models of climate, we shall discuss here some of the mechanisms of importance.

OCEAN-ATMOSPHERE COUPLING

There is little doubt that the oceans affect the climate in a very important way. For example, they transport and store large amounts of heat and thus reduce the latitudinal gradient of temperature and the amplitude of seasonal temperature variation in the atmosphere. Moreover, they supply water vapor to the atmosphere and control the hydrologic cycle. Since the oceans are so closely coupled to the atmosphere, any change in climate usually accompanies the change in sea surface temperature, and vice versa (see Section 7.3). Thus it is necessary to incorporate the effects of the ocean-atmosphere coupling into numerical models in order to evaluate the possibility of climatic change.

A preliminary version of such a model of the joint ocean-atmosphere system was constructed recently by Manabe and Bryan (1969). Figure 6.10 shows how the atmospheric part of their model interacts with the oceanic part. As this figure indicates, heat is exchanged through radiative fluxes and turbulent transport of sensible and latent heat. Water is exchanged through precipitation and evaporation, and momentum is exchanged through surface wind stress at the oceanic surface. The distribution of zonal mean

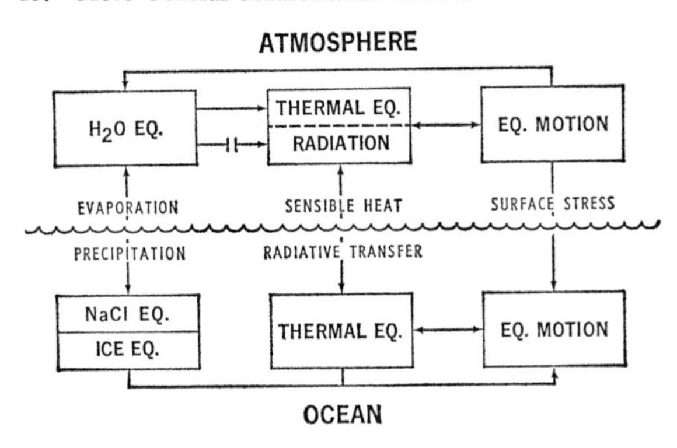

Figure 6.10 Block diagram of the joint model structure of Manabe and Bryan, 1969.
Source: Manabe, 1969b.

temperature obtained from the time integration of the model is shown in Figure 6.11 as an illustration of the results of the computation. Unfortunately, this model does not contain some of the important feedback mechanisms discussed in the following parts of this subsection and in Section 6.5. It also has a limited computational domain and highly idealized topography. The construction of global models of the joint ocean-atmosphere system with the feedback mechanism to be described is necessary for the evaluation of the possibility of climatic change.

WATER-VAPOR–"GREENHOUSE" COUPLING

It is known that, given sufficient time, the atmosphere tends to attain a definite climatological distribution of relative humidity in response to a change of temperature (see Sections 6.4.2 and 6.5.2). This restoring tendency is responsible for a positive feedback mechanism which may be called "greenhouse water vapor coupling." It involves the following chain processes: warmer temperature→more water vapor in the atmosphere→more "greenhouse" effect→warmer temperature. In their study of radiative, convective equilibrium, Manabe and Wetherald (1967) show that this mechanism approximately doubles the sensitivity of equilibrium temperature to a change of solar constant. In some numerical models of the atmosphere, this feedback mechanism is nullified by using the climatological distribution of water vapor

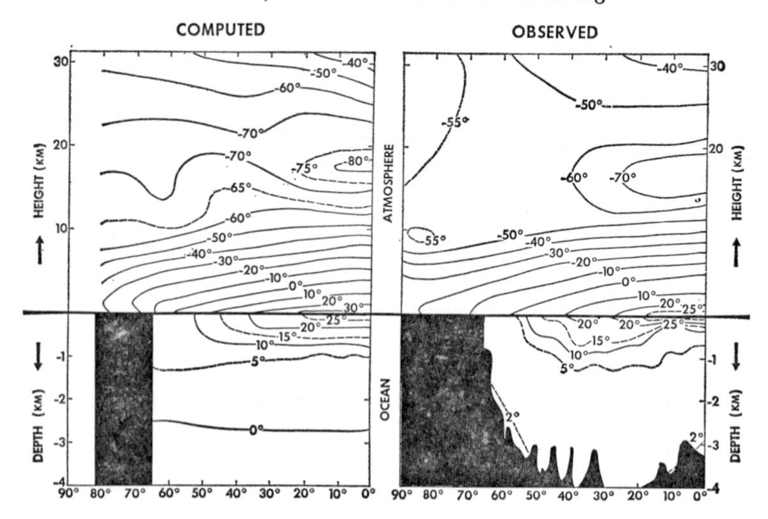

Figure 6.11 Left-hand side: zonal mean temperature of the joint ocean-atmosphere system. Right-hand side: observed distribution in the Northern Hemisphere.
Source: Manabe, 1971.

for the computation of radiative transfer, rather than the distribution obtained from the prognostic equation for water vapor (refer to Figure 6.7). For the reasons mentioned earlier it is necessary to incorporate this effect into models for the study of climatic change.

SNOW-COVER–ALBEDO–TEMPERATURE COUPLING

Since the albedo of snow or ice pack is much larger than that of bare soil or ocean, the temperature in higher latitudes is strongly controlled by the distributions of snow cover and ice pack (see Sections 6.5.3 and 6.7). This coupling involves the following positive feedback processes: wider snow cover→more reflection of insolation→colder surface temperature→wider snow cover. As suggested by Budyko (1969), this mechanism may have very important effects upon the sensitivity of climate, particularly in high latitudes. In order to study the sensitivity of climate, it is essential to incorporate the prognostic scheme of snow cover and ice pack in a mathematical model of the joint ocean-atmosphere system.

SEASONAL VARIATION OF INSOLATION AND SNOW COVER

Since the depth of the snow cover (ice pack) increases during the winter and decreases during the summer, seasonal variation of solar radiation has strong controlling effects over the net change of snow depth from one year to the next. Therefore, it is important to incorporate the effects of the seasonal variation of solar radiation in order to discuss the stability of the snow cover (ice pack) or the sensitivity of climate.

CLOUD-COVER–RADIATION-BALANCE COUPLING

The importance of cloud cover upon the radiation balance of the earth-atmosphere system was discussed earlier in Sections 5.2.1, 5.2.2, and 6.5.4. In order to incorporate the effect of this coupling into a mathematical model of the atmosphere, it is necessary to determine the distribution of clouds as a function of the large-scale distribution of quantities such as relative humidity, temperature, and vertical velocity. As we shall point out in the following section, comprehensive schemes of determining cloudiness have not yet been developed. For the study of climatic change, incorporation of this coupling into numerical models of the atmosphere is important.

6.8.3
Outstanding Problems

IMPROVEMENT OF ATMOSPHERIC MODELS

Improvement of mathematical models of the atmosphere is one of the major objectives of the Global Atmospheric Research Program (GARP), and comprehensive plans for the improvement of models have already been proposed. Here we shall limit ourselves to discussion of improvements that are particularly desirable for the study of climatic change.

The numerical time integration of the system of prognostic equations on a spherical surface is not a trivial task. Each of the various global grid systems that have been developed so far has its own difficulties. For example, the global grid system proposed by Kurihara tends to develop an excessively strong polar high (Kurihara and Holloway, 1967). A global grid system and accompanying finite difference scheme free of these computational difficulties has yet to be proposed (Phillips, 1971). Therefore the development of a global grid system and accompanying finite-difference scheme free of these difficulties is very urgent for the successful simulation of climate.

Another area of modeling that needs further refinement is the treatment of clouds. Clouds have a very strong influence on the field of radiative transfer and may have significant effects upon the way in which the climate changes (see Section 5.2 and 6.5.4). In order to incorporate the cloud-radiation balance coupling, it is necessary to determine cloudiness as a function of the fundamental variables of a mathematical model. In their model, Washington and Kasahara (1970) determined the cloud distribution from the sign of the vertical velocity. On the other hand Smagorinsky (1960) attempted to relate the amounts of nonconvective cloud to the large-scale distribution of relative humidity. A comprehensive scheme of cloud prediction that accurately predicts the large-scale distribution of clouds must be developed and incorporated into the models. Observation of clouds from satellites at visible and infrared wavelengths may yield very useful information for this purpose.

For an accurate estimate of the effect of clouds on the radiation field, it is also necessary to know the optical properties of clouds. Both observational and theoretical studies on the optical properties of clouds should be used. Satellite observation of net incoming solar radiation and net outgoing terrestrial radiation can be useful in the verification of both the scheme of cloud prediction and the computation of radiative transfer in a cloudy atmosphere.

In most of the schemes for computing radiative transfer that have so far been incorporated in mathematical models of the atmosphere, the effect of particles is not considered. The effect of particles should be incorporated into the models so that one can discuss the effect of particles on climate and, in particular, the effect of man-made particles on climate change. In order to do this, more information on the optical properties of aerosols is needed (see Section 8.7).

A simulation of the hydrological cycle in the models needs improvement. As noted before, the distribution of hydrologic elements such as snow cover and soil moisture at the earth's surface strongly influences climate. The hydrologic scheme that is incorporated into the existing model of climate is highly idealized for the sake of simplicity (for example see Manabe, 1969 a, b). It is necessary to verify the prognostic scheme of snow cover and soil moisture against observation and to make necessary improvements of the scheme.

Improvement in the modeling of moist convection is an important task. One of the major difficulties in constructing a numerical model of the atmosphere comes from our ignorance of the effects of moist convection on the large-scale distributions of temperature, moisture, and momentum. In order to develop a better method of parameterizing these effects, extensive observational studies are being carried out under the Global Atmospheric Research Program (GARP). Theoretical studies of the statistical dynamics of moist convection, such as those carried out by Asai and Kasahara (1966), are strongly encouraged. It is hoped that one can also develop a practical method of predicting the distributions of convective clouds by using the results from these observational and theoretical studies.

IMPROVEMENT OF OCEAN MODELS

Numerical modeling of the ocean circulation has been attempted by Sarkysian (1962), Bryan and Cox (1967), and others. They have been able to simulate such fundamental features of oceans as the Gulf Stream and the general character and depth of the thermocline. However, the mean temperature and salinity of a large fraction of the ocean are now known in fair detail from careful observations taken over many years. An adequate joint atmosphere-ocean model should be able to reproduce several important features of oceanic climate. This ability may prove to be a fairly critical test.

Physical oceanographic theory describes with some success the depth-averaged transport of water mass induced by winds blowing over the ocean. The transport of heat, however, generally depends not only upon water transport and on the total heat content but on the distribution of both with depth. These distributions depend critically on the vertical mixing processes in the upper layers of the ocean-processes that are very poorly understood. Increased emphasis upon coordinated observational and theoretical studies of vertical mixing processes in the upper layer of the ocean should be a focal point for further research.

Even when a model closely reproduces contemporary temperature and salinity distributions, there will remain some doubt about its suitability for dealing with changes in climate. This doubt will be removed only by demonstrating the success of the model in reproducing temporal changes on some suitable time scale. The task is an enormous one, and the number of scientists working in this field, a much smaller number than are working on atmospheric models, is inadequate to the task.

A realistic model of the ocean can be attained only by close interaction between the modelers and the observational oceanographers, because, unlike the situation in meteorology, there is no regularly collected extensive body of oceanographic data against which time-dependent models may be tested, nor is there now any generally agreed-upon prescription for obtaining suitable data. Data gathering in oceanography is extremely expensive, so that any extensive program must be very carefully designed. Thus more scientists should be encouraged to participate in ocean-modeling research, and modelers and observational oceanographers should work closely together for the development of observational programs capable of generating the data necessary for the verification and testing models.

The prediction of sea ice formation is an important remaining step in improving ocean models. In the preceding section, it was pointed out that the ice pack has a very large reflectivity and can strongly control the climate of higher latitudes. Therefore, it is essential to incorporate the prognostic scheme for the ice pack in mathematical models of the ocean intended for climatological studies. Extensive observational and theoretical studies should be encouraged for the development of prognostic schemes which can predict with sufficient accuracy the extent and thickness of sea ice.

6.8.4
Computer Time

One of the factors that should be taken into consideration for any modeling research is the computer time required for carrying out the numerical computation. The experience gained from numerical time integrations of an atmospheric or an oceanic model indicate that, starting from arbitrary initial conditions, it takes less than 1 model year for the thermal structure of the amosphere to approach a quasi-steady state; whereas it takes more than 100 model years for the ocean to "settle down." Accordingly, the thermal relaxation time of the joint model should be the longer of the two, that is more than 100 years. In Table 6.2 the computer time for the IBM 360-91 (which is used for the time integration of the atmospheric model and the oceanic model developed at the Geophysical Fluid Dynamics Laboratory of the National Oceanic and Atmospheric Administration [NOAA], Princeton), is listed for two different computational resolutions. This table indicates that the computer time required for reaching a steady model climate is very long indeed. Although it is projected that a computer one or two orders of magnitude faster than IBM 360-91 will become available soon, it is highly desirable to devise methods by which computer time can be minimized.

One example of such a scheme of minimization may be found in a study by Manabe and Bryan (1969). In their study, a 1-year integration of the atmospheric part of the model was performed concurrently with a 100-year integration of the oceanic part of the model. In this way, both parts of the model were expected to approach the final state with optimum speed.

Another possibility for reducing the required computer time is to reduce spatial resolution. (Note that, by doubling mesh size, one can reduce machine time by a factor of approximately 8.) As demonstrated by Manabe et al. (1970), the model climate becomes

Table 6.2
Machine Time Required for the 1-year Integration of the Atmospheric and the Oceanic Model with Global Domain by the IBM 360-91

	Number of Vertical Finite-Difference Levels	Grid Size (km)	IBM 360-91 Machine Time (hr/yr)
Atmospheric model	9	250	960
Atmospheric model	9	500	120
Oceanic model	9	220	60
Oceanic model	9	440	8

Source: Manabe, 1971, private communication.

less realistic if the grid size is increased from 250 kilometers to 500 kilometers; however, the model with coarse computational resolution may be good enough for investigation of certain aspects of climatic changes.

There are other possibilities for economizing. Although it is probably necessary to take into consideration the full depth of the ocean when dealing with climate change, it may be possible to discuss short-period changes of climate by use of a joint model with a shallower ocean and thus drastically reduce the thermal relaxation time of the ocean.

These are simply a few obvious ways of minimizing the computer time. Although none of them seems to be completely satisfactory, they are listed here in order to stimulate more and better ideas. Better methods are clearly needed for reaching a climatic equilibrium in a minimum length of computer time.

6.8.5 Numerical Experimentation

Once a mathematical model of climate is completely formulated, it is possible to perform various numerical experiments with the model. Some of the major objectives of such experiments are the following:

1. Ascertain the similarities of the model to the actual atmosphere (simulation experiment).
2. Improve our understanding of climate (controlled experiment).
3. Investigate the change of climate due to man's activity (SMIC experiment).

SIMULATION EXPERIMENT

Before discussing the possibility of climatic change, it is essential to ascertain that the behavior of the model is similar to that of the actual atmosphere. One possible method of verification is to simulate the seasonal variation of climate with a numerical model of climate. Since the seasonal variation is the largest observable climatic change, we believe this is one of the best methods of ascertaining the similarity of the model to the actual atmosphere.

An excellent method of testing atmospheric models is to use them for the purpose of numerical weather prediction and check their performance against the actual change of weather. Extensive numerical experiments of this kind are being undertaken under the Global Atmospheric Research Program. As pointed out in the preceding section, similar verification of the transient behavior of ocean models (or joint models) should be strongly encouraged.

CONTROLLED EXPERIMENT

It is desirable to understand how climate is determined before attempting to evaluate the possibility of climatic change. One can gain a general understanding of climate by identifying the role of the major factors that control the climate. In order to do this, one can compare the results from two numerical experiments with and without the relevant factor or with variations of that factor. For example, the effects of seasonal variation of solar radiation on climate can be isolated by comparing two numerical experiments of the model with and without seasonal variation of insolation.

The integration of a general model of climate would obviously consume an enormous amount of computer time but is not required for all experiments. One can carry out a variety of numerical experiments using idealized models with varying degrees of simplification and can gain a deeper insight into the mechanism of climate determination and change (see Sections 6.4, 6.5, 6.6, and 6.7).

SMIC EXPERIMENT

When a realistic model of the joint ocean-atmosphere system is completely formulated, one can then test the possibility of climatic change due to human activity. For example, one can evaluate the climatic change resulting from a change in CO_2 content in the atmosphere by comparing the climates of two models with different CO_2 contents. Or one can evaluate the magnitude of the change in regional climate due to thermal pollution by introducing local heat sources in certain regions of the model continent (see, for example, Washington, 1971). It seems to be good strategy to evaluate first the effects of various factors separately and then to evaluate the integrated effect.

6.8.6
Recommendations

1. We recommend the construction of joint ocean-atmosphere models with global scale for the study of climatic change. This will require acceleration of ocean modeling research. All major feed-

back mechanisms should be incorporated into this model so that the sensitivity of the model climate is realistic. The joint model should be verified through attempts to simulate the seasonal variation of climate. This model could be used effectively to evaluate man's impact on the climate.

2. We recommend improvement of the global grid system and accompanying finite-difference scheme by which we can perform the time integration of the prognostic equations.

3. We recommend development of a comprehensive scheme of cloud prediction that accurately predicts the large-scale distribution of clouds. Estimates of the optical properties of clouds must also be improved if the model is to be realistic.

4. We recommend investigation of the optical properties of particles and incorporation of the radiative effects of aerosols into numerical models of the atmosphere.

5. We recommend improvement of the prognostic scheme of soil moisture and snow and ice cover incorporated in the model.

6. We recommend seeking various methods of minimizing computer time required for reaching climatic equilibrium.

6.9
General Conclusions

Climate is determined by a balance among numerous interacting physical processes in the oceans and atmosphere and at the land surface. Locally and globally, climate is subject to change on all scales of time, but the physical processes themselves remain the same and are amenable to study by statistical, physical, and mathematical techniques. If we are to assess the possibility and nature of a man-made climatic change, we must understand how the physical processes produce the present climate and also how past changes of climate, clearly not man-made, have occurred.

Several physical-mathematical techniques—they have come to be called "models"—are being developed to attack the problem. We have distinguished four types: (1) "global-average" models, in which horizontal motion of the atmosphere is neglected; (2) parameterized semiempirical models that consider the whole atmosphere and surface but simulate some of the effects of atmosphere and oceanic motions with the aid of empirically adjusted constants; (3) statistical-dynamical models in which physical laws are applied to statistics of the atmospheric variables; and (4) explicit numerical models in which motions and interactions are treated in detail by integrating mathematical equations expressing the time rate of change of the variables, though in practice the detailed treatment must be abandoned at some minimum scale of motion and replaced by empirical parameterization or statistical methods.

The global-average models serve the purpose of identifying and investigating in a preliminary way problem areas of concern to climate theory; for example, they have been used to make first estimates of the effects of changing CO_2 content. The semiempirical models allow consideration of more complex problems of climate, such as the stability of the polar ice caps. There may be some uncertainty about their indications because we do not understand in detail the parameterizing processes or the consequences of parameterization; nevertheless, their predictions must not be ignored, particularly if, as is often the case, the indications of independently developed models are similar. Statistical-dynamical models are in an early stage of development, but it is hoped that they will at some later stage obviate some of the enormous and costly computational load associated with explicit numerical models. Significant progress has been made in developing these detailed models, enough to give appreciation of the magnitude of the task and confidence that it might succeed.

We now know enough of the theory of climate and the construction of climatic models to recognize the possibility of man-made climatic change and to have some confidence in our ability ultimately to compute its magnitude. Recent results obtained using empirical models have in our view increased the urgency of the study of climate theory in its own right. Some of these results, for example, concern the delicate balance of the processes that maintain the arctic sea ice. The empirical models suggest that a small change in mean air temperature or in solar radiation reaching the surface could result in considerable expansion or contraction of the ice pack—a climatic change of great significance to human life. This in our opinion reinforces the urgency of studies of climate by all available methods in order to understand its natural and possible man-made changes. Some financial implications of these recommendations are discussed in Section 1.4.

References

Asai, T., and Kasahara, A., 1966. A theoretical study of the compensating downward motions associated with cumulus clouds, *Journal of Atmospheric Sciences, 24:* 487–496.

Blinova, Ye. N., 1943. A hydrodynamical theory of pressure and temperature waves and of centers of atmospheric action, *Doklady Akademii Nauk SSSR, 92:* 557–560.

Brooks, C. E. P., 1950. *Climate through the Ages: A Study of the Climatic Factors and Their Variations* (New York and Toronto: McGraw-Hill).

Bryan, K., and Cox, M. D., 1967. A numerical investigation of the oceanic general circulation, *Tellus, 19:* 54–80.

Budyko, M. I., 1961. Heat and water balance theory of the earth's surface, the general theory of physical geography and the problem of the transformation of nature, Water-Heat Balance Symposium of the 3rd Congress of the Geographical Society of the U.S.S.R., p. 2118.

Budyko, M. I., 1963. *The Heat Budget of the Earth* (Leningrad: Hydrological Publishing House).

Budyko, M. I., 1969. The effect of solar radiation variations on the climate of the earth, *Tellus, 21:* 611–619.

Budyko, M. I., 1971. *Climate and Life* (Leningrad: Hydrological Publishing House).

Budyko, M. F., and Vasischeva, 1971. (In press).

Donn, W., and Shaw, D., 1966. The heat budgets of an ice-free and an ice-covered Arctic Ocean, *Journal of Geophysical Research, 71:* 1087–1093.

Fletcher, J. O., 1966. The arctic heat budget and atmospheric circulation, *Proceedings of the Symposium on the Heat Budget and Atmospheric Circulation,* RAND Corporation, Santa Monica, California.

Green, J. S. A., 1970. Transfer properties of the large-scale eddies and the general circulation of the atmosphere, *Quarterly Journal of the Royal Meteorological Society, 96:* 157–185.

Haurwitz, B., 1948. Insolation in relation to cloud type, *Journal of Meteorology, 5,* 110–113.

Holloway, J. L., Jr., and Manabe, S., 1971. Simulation of climate by a global general circulation model: I. Hydrologic cycle and heat balance, *Monthly Weather Review, 99:* 335–370.

Kasahara, A., and Washington, W. M., 1967. NCAR global general circulation model of the atmosphere, *Monthly Weather Review, 95:* 389–402.

Kraus, E. B., and Lorenz, E. N., 1966. Numerical experiments with large-scale seasonal forcing, *Journal of Atmospheric Sciences, 23:* 3–12.

Kurihara, Y., and Holloway, J. L., Jr., 1967. Numerical integration of a nine-level global primitive equation model formulated by the box method, *Monthly Weather Review, 95:* 509–530.

Leith, C. E., 1965. Numerical simulation of the earth's atmosphere, *Methods in Computational Physics,* 4 (New York: Academic Press), pp. 1–28.

London, J., 1957. A study of the atmospheric heat balance (New York University College of Engineering, Research Division, Oceanography), Final Report on Contract, No. AF 19 (122)–165.

London, J., and Sasamori, T., 1971. Radiative energy budget of the atmosphere, *Man's Impact on the Climate,* edited by W. H. Matthews, W. W. Kellogg, and G. D. Robinson (Cambridge, Massachusetts: The M.I.T. Press), pp. 141–155.

Lorenz, E. N., 1965. A study of the predictability of 28-variable atmospheric model, *Tellus, 17:* 321–333.

Lorenz, E. N., 1968. Climatic determinism, *Meteorological Monographs, 8:* 30, 1–3.

Manabe, S., 1968. *Meteorological Monographs,* Vol. 8, No. 30, p. 28.

Manabe, S., 1969a. Climate and ocean circulation. Part I: The atmosphere circulation and the hydrology of the earth's surface, *Monthly Weather Review, 97:* 739–774.

Manabe, S., 1969b. Climate and ocean circulation. Part II. The atmospheric circulation and the effect of heat transfer by ocean currents, *Monthly Weather Review, 97:* 775.

Manabe, S., 1970. Cloudiness and the radiative convective equilibrium, *Global Effects of Environmental Pollution,* edited by S. F. Singer (Dordrecht, Holland: Reidel Publishing Company; New York: Springer-Verlag), pp. 156–157.

Manabe, S., 1971. Estimates of future change of climate due to the increase of carbon dioxide concentration in the air, *Man's Impact on the Climate,* edited by W. H. Matthews, W. W. Kellogg, and G. D. Robinson (Cambridge, Massachusetts: The M.I.T. Press), pp. 249–264.

Manabe, S., and Bryan, K., 1969. Climate calculation with a combined ocean-atmosphere model, *Journal of Atmospheric Sciences, 26:* 786–789.

Manabe, S., Smagorinsky, J., Holloway, J. L., Jr., and Stone, H. M., 1970: Simulated climatology of a general circulation model with a hydrologic cycle, Part 3, Effect of increased horizontal computational resolution, *Monthly Weather Review, 98:* 175–212.

Manabe, S., Smagorinsky, J., and Strickler, R. F., 1965. Physical climatology of a general circulation model with a hydrologic cycle, *Monthly Weather Review, 93:* 769–798.

Manabe, S., and Wetherald, R. T., 1967. Thermal equilibrium of the atmosphere with a given distribution of relatively humidity, *Journal of Atmospheric Sciences, 24:* 241–259.

Mintz, Y., 1965. Very long term global integration of the primitive equation of atmospheric motion (WMO-IUGG Symposium on Research and Development Aspects of Long Range Forecasting, Boulder, Colorado, 1964, and Geneva, 1965), WMO Technical Note 66, pp. 141–161.

Möller, F., 1951. Vierteljahrskarten des Niederschlags für die Ganze Erde (Quarterly charts of rainfall for the whole earth), *Petermanns Geographische Mitteilungen,* Vol. 95, No. 1 (Gotha, Democratic Republic of Germany: Justus Perthes), pp. 1–7.

Möller, F., 1963. On the influence of changes in the CO_2 concentration in air on the radiation balance of the earth's surface and on the climate, *Journal of Geophysical Research, 68:* 3877–3886.

Phillips, N. A., 1956. The general circulation of the atmosphere: A numerical experiment, *Quarterly Journal of the Royal Meteorological Society, 82:* 123–164.

Phillips, N. A., 1971. Numerical weather prediction, *EOS (Transactions of the American Geophysical Union).*

Rakipova, L. R. 1962. Climate change when influencing the arctic basin ice (in Russian), *Meteorologiia I Gidrologiia, 9.*

Rasool, S. I., and Schneider, S. H., 1971. Atmospheric carbon dioxide and aerosols: effects of large increases on global climate, *Science, 173:* 138–141.

Saltzman, B., 1968. Steady state solutions for axially symmetric climatic variables, *Pure and Applied Geophysics, 69:* 237–259.

Sarkisyan, A. S., 1962. Odinamike vozniknoveniia vetrovykh techeniĭ v baroklinnom okeane (On the dynamics of the origin of wind currents in a baroclinic ocean), *Okeanologiia, 2:* 393–409.

Sellers, W. D., 1969. A global climatic model based on the energy balance of the earth-atmosphere system, *Journal of Applied Meteorology, 8:* 392.

Smagorinsky, J., 1960. On the dynamical prediction of large-scale condensation by numerical method (American Geophysical Union), Monograph No. 5, pp. 71–78.

Smagorinsky, J., 1963. General circulation experiments with the primitive equations, 1, The basic experiment, *Monthly Weather Review, 93:* 99–164.

Smagorinsky, J., Manabe. S., and Holloway, J. L., Jr., 1965. Numerical results from a nine-level general circulation model of the atmosphere, *Monthly Weather Review, 93:* 727–768.

Washington, W. M., 1971. On the possible use of global atmospheric model for the study of air and thermal pollution, *Man's Impact on the Climate,* edited by W. H. Matthews, W. W. Kellogg, and G. D. Robinson (Cambridge, Massachusetts: The M.I.T. Press), pp. 265–276.

Washington, W. M., and Kasahara, A., 1970. A January simulation experiment with the two-layer version of the NCAR global circulation model, *Monthly Weather Review, 98,* 559:580.

Yamamoto, G., and Tanaka, M., 1971. Increase of global albedo due to air pollution (unpublished).

Part IV Effects and Impacts

7 Climatic Effects of Man-made Surface Change

7.1 Modification of the Microclimate

7.1.1 Discussion

There is no doubt that whenever man changes the landscape he modifies the microclimate. Ryd (1970) has used the phrase "climatological sheath" to describe the space around a single building, within which there are rain shadows and anomalies in wind, temperature, humidity, and soil moisture (Figure 7.1). Similar sheaths surround other local landscape features (natural and man-made) such as ponds, trees, and ploughed fields. The effects on this microscale have been studied for many years and are reasonably well understood.

When groups of structures merge into towns and cities, the term "climatological dome" is used to describe the phenomenon illustrated in Figure 7.2; within this dome there are well-documented meteorological anomalies (see, for example, Peterson, 1969). The microclimates of urban parks and on the shaded sides of streets remain as special anomalies, but on the larger scale of the city many common features can be detected, sometimes up to heights of a kilometer. Frequently the top of the dome (the so-called *urban mixing height*) is well defined and can be identified quite readily from an aircraft.

Large rural areas that have been greatly modified by irrigation or deforestation also have climatological domes, although the top of the mixed layer may not be so evident visually as it is over cities. In all such cases (including the city), the main features are reasonably well understood and can in fact be modeled numerically with some success. Although we do not wish to consider this scale in detail, two particular features should be mentioned by way of introduction to the discussion of regional and global effects.

First, an urban thermal and pollution plume exists whenever a wind is blowing, transporting heat and matter downstream out of the city and modifying, for example, the rural radiation balance for at least a few kilometers. The urban plume has been studied by Clarke (1969) in Cincinnati and by Oke and East (1971) in Montreal and is of the schematic form shown in Figure 7.3. On this scale, therefore, man-induced influences may begin to have an effect on a significant fraction of the surrounding countryside.

The second feature of interest occurs whenever the regional flow is weak. Thermal gradients set the air in motion and urban wind cells develop as illustrated schematically in Figure 7.4. On the microscale too, very local convective cells may develop, but the air motions are then too small to be detected easily. Even on the scale of a city, the inflow near ground level and the outflow aloft are difficult to measure. At night particularly, inward drift of surface air is intermittent and irregular, due to the obstructions in the built-up area.

In both the steady- and the light-wind cases, numerical modeling of urban dispersion is proving to be a useful tool that may soon provide sufficiently reliable predictions to be of value for the decision makers. In most models, in order to avoid the loss of signal resulting from the use of long averaging times, predictions of transport and dispersion patterns are made hourly. The first step is to obtain a detailed inventory of pollution source strengths, city block by city block. Next, the meteorological processes are parameterized, and diffusion equations are solved for each source. By combining the results, the areal patterns can be obtained for a particular hour. The next step is to introduce some simple weather classification scheme such as that of Pasquill (1962) to obtain climatological air-quality frequency distributions from long-term meteorological observations. Once the model has been "validated" by comparison with real data, it can be used to interpolate between sampling points and to simulate the effects of changing land-use patterns (Stern, 1971). This approach could readily be adapted to rural modifications (patchy irrigation, and so on) of the same dimensions as a city, although no studies of this type have been reported as yet.

Before turning to the regional scale, we should note also that the effects of surface anomalies depend on the latitude, aridity, and the like, that is, on the climatic zone within which the man-made disturbance is introduced. Over subarctic cities during win-

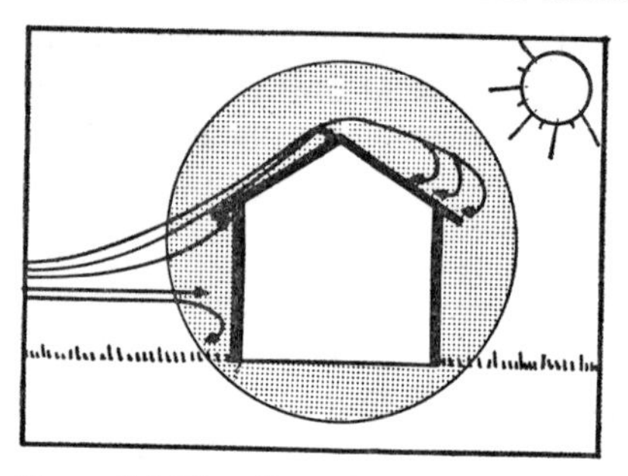

Figure 7.1 The climatological sheath around a structure.
Source: Ryd, 1970.

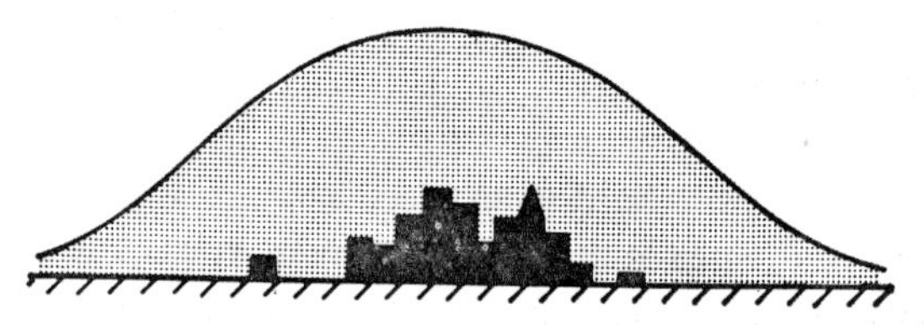

Figure 7.2 The urban dome.

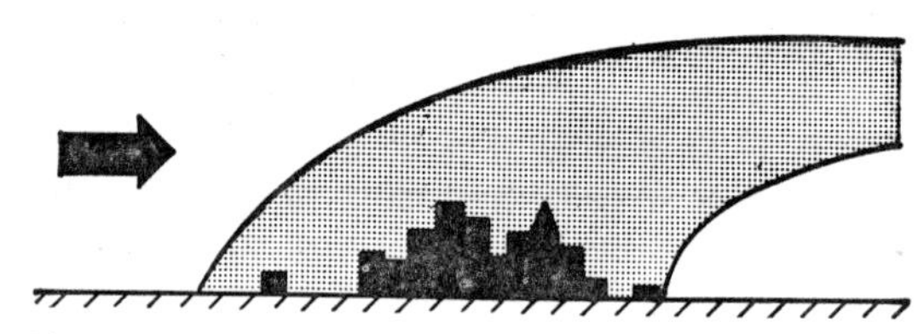

Figure 7.3 The urban thermal and pollution plume occurring when a regional wind is blowing.

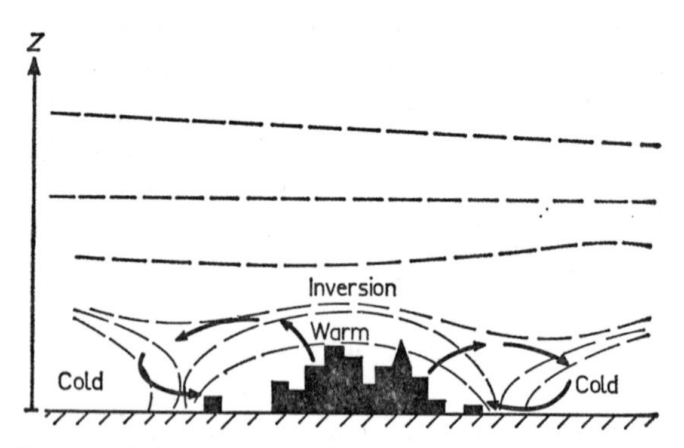

Figure 7.4 The urban circulation developing when regional winds are light.
Source: Landsberg, 1970.

ter when winds are light, for example, the thickness of the urban dome is rather small, not exceeding 100 meters in many cases. An oasis in the subtropics, on the other hand, is surrounded by a mixed layer several kilometers in vertical extent, and upward diffusion of water vapor is so rapid that despite the fact that the evapotranspiration rate is very high in the oasis there is no significant rise in air humidity. We need only recall that the relative humidity at shipboard level in the West Indies averages about 85 percent (rather than 100 percent), although the air has been over the ocean surface for many days. This effect is due not only to the turbulent mixing in the air below the trade-wind inversion but also to the boundary-layer resistance at the air-sea interface itself. Because turbulence is suppressed near a surface, the exchange of saturated with unsaturated air within a few millimeters of the ocean is retarded. Similar boundary-layer resistances occur over land, slowing down the exchange of water, small particles, CO_2, and other trace gases. Thus, for example, vegetation ceases to be an abundant source for water vapor and a sink for CO_2 when the stomata are closed.

Turning now to the regional scale (approximately 1000 km), the atmospheric dispersion models designed to test the effect of man-made regional anomalies are not yet well formulated, even on the time scale of a few days during carefully selected weather patterns. There are several reasons for this, as discussed by Munn and Bolin (1971). Probably the most important factor, however, is that the dome of influence now extends well up into the troposphere, into a region that is not well observed in terms of the parameters required for application of the classical diffusion equations. From satellite photographs, we can observe the drift of haze from the eastern United States to the central Atlantic and the drift of dust from West Africa to the Caribbean. The vertical dimensions of these clouds remain to be investigated, however. In most regional studies, therefore, the assumption is made that the troposphere is a well-mixed reservoir, although the modelers recognize that sharp discontinuities exist on time scales of a few days in the temperate zones and on time scales of a few weeks in the trade-wind belts. The urban numerical models can in principle be modified for use on the regional scale, but we must obtain much more information than is available at present on the properties of the tropospheric "dome," particularly by a regular program of aircraft monitoring.

In the subsections to follow, we implicitly accept the concept that man is modifying the climate on the microscale and within areas the sizes of cities. We shall therefore consider the regional climatic influences of man, and we shall also speculate on the effects of regional anomalies on global climate. What must be the dimensions of a megalopolis or of a glacier in order to affect significantly the general circulation of the atmosphere? These are questions that cannot readily be answered, but we believe that they must be asked, in the context of world climate.

Finally, we make a rather general recommendation. Those who wish to follow climatic trends have no difficulty in obtaining data for individual meteorological stations. However, each investigator must compute his own regional and global spatial averages, a time-consuming process. Because ways of determining spatial averages are rather subjective, a comparison of results obtained by different scientists is difficult.

7.1.2
Recommendation

We recommend that an appropriate international agency be invited to consider ways of standardizing methods for computing seasonal regional averages of surface temperature, precipitation, and radiation (or cloudiness), and that the agency be invited subsequently to facilitate the calculation of such averages. We believe that areal averages of meteorological surface elements averages are valuable indicators of climatic change though we recognize that such averages are biased because of varying local exposure and grid densities.

7.2
Snow and Ice Cover

Climatological estimates of surface albedo (ratio of reflected to incident solar radiation) are given in Table 7.1 (Budyko, 1971). These estimates show that there is a change in the surface radiation balance, and thus in the surface heat balance, whenever there is a man-induced manipulation of the snow and ice cover. Dusting

Table 7.1
Climatological Estimates of Surface Albedo

Nature of Surface	Albedo
Stable snow cover (latitudes above 60°)	0.80
Stable snow cover (latitudes below 60°)	0.70
Forest with stable snow cover	0.45
Forest in spring with unstable snow cover	0.38
Steppe and coniferous forests in summer	0.13
Deciduous forests in summer	0.18
Oceans, latitude 70°	0.23–0.09
Oceans, latitude 60°	0.20–0.07

Source: Budyko, 1971.

of the surface with soot has been shown to accelerate greatly the spring melt over small experimental areas, particularly in polar latitudes, by lowering the albedo substantially (Arnold, 1961). It is possible that the arctic sea ice could be largely removed by a massive aircraft dusting program during one or a few summers when the atmospheric radiation balance was relatively favorable. However, it is not clear whether the ice would re-form immediately during the following winters (see also Section 6.7). In this section we shall discuss this question from the meteorological point of view, postponing the oceanographic consideration until later (Section 7.3.1). On the one hand, cold-air advection from the neighboring continents would greatly increase convective and evaporative heat losses and would generate snow squalls, as in the lee of the Great Lakes, increasing precipitation and permitting frozen areas to expand. On the other hand, the resulting cloudiness as well as the modified surface temperature (about −2°C rather than perhaps −20°C) would reduce the surface net long-wave radiation losses, removing a principal cause of cooling. In addition, the increased snow cover would further insulate the underlying ice and would reduce its winter growth rate.

The state of the arctic sea ice is a critical factor in determining the climate of the entire Northern Hemisphere because the winter semipermanent polar anticyclone provides one of the anchors for the general circulation. Simple, theoretical estimates (Flohn, 1964), supported by empirical data (Flohn and Korff, 1969), seem to indicate that with a decrease of the temperature difference between equator and pole the subtropical anticyclone belt would be displaced northward. In the subarctic and temperate zones, man is increasingly affecting the ice cover over large lakes, inland seas, and gulfs. These effects are the results of the soot that is deposited from adjacent industrial areas and also of the use of ice breakers. These activities accelerate the spring thaw and delay the autumn freeze. In winter, however, open leads may actually permit the water to cool more rapidly and thus may cause the surrounding ice to thicken. We think, however, that although ice breakers may have some marginal effects on local climate, the influence of gales that shift and crack the ice pack is at least an order of magnitude greater than that caused by man. The arctic sea ice is discussed in more detail in Section 7.3.

Snow removal in cities, together with the modifying influence of soot deposited on undisturbed snow patches in parks and gardens, contributes substantially to the urban heat-island effect. On both the city and the regional megalopolis scales, the radiation balance and the surface temperature field are modified by these man-made effects within a day or so of each new snowfall.

In nonurban areas a change from snowcovered forest to snowcovered grassland, or vice versa, makes a significant difference in the surface radiation balance. In winter, because the incoming solar radiation is not great, the climatic modification is slight. In spring, however, the runoff is delayed in the forest, and the region may remain "in cold storage" for a few weeks longer than if the land were cleared.

In the temperate latitudes the main surface inputs to the general circulation are through the process of cold-air advection over the coastal waters just east of Asia and of North America. We speculate that a snow-cover anomaly extending across half a continent, due either to man-made manipulation of the surface cover or to an unusual succession of weather events, may affect the atmospheric general circulation on a time scale of the same order of magnitude as that of the lifetime of the anomaly. We believe, however, that this will be a purely transient perturbation with no demonstrable effect on long-term climate.

The mean summer surface position of the arctic front corresponds to the tree line, the vegetation presumably integrating the effects of all the climatic elements, but with a time lag of perhaps a century (Hare, 1951; Bryson, 1966). We can only speculate about the effect of a natural or man-made change (by forest fires or cutting) on the position of the tree line, but we believe that the influence will be insignificant except on the microscale. However, we recommend that ecoclimatological studies be encouraged, not only in connection with the arctic tree line but more generally throughout all the climatological zones of the world.

7.3 The Oceans

As has been repeatedly pointed out in earlier chapters (see Sections 5.3 and 6.8, for example), the ocean exerts a very great influence on climate. It stores and transports heat and is the major source of atmospheric water vapor, and it is certainly not immune to influence from man's activities.

7.3.1 Removal of the Arctic Sea Ice

As was pointed out in Section 7.2, there seems to be little doubt that arctic sea ice could be eliminated by decreasing the albedo by dusting or by some other deliberate intervention (Donn and Shaw, 1966). In the former case, there is some doubt as to whether the transition to an open ocean would persist. However, it appears very probable that the arctic sea-ice cover is in a very sensitive state and that rather small changes in any of a number of parameters influencing that ocean could lead to substantial changes in the extent of the ice. Such parameters include changes in the transport of Atlantic water into the Arctic, changes in the amount and timing of river flows, and especially changes in the overall temperature regime at high latitudes.

There is a distinct possibility—according to some a probability—that a temperature rise associated with the anticipated injections of heat and CO_2 into the atmosphere in the next century would result in the summer melting of arctic ice (see Section 6.7).

The mean lifetime of arctic sea ice is less than 10 years, and it is possible that the transition from an ice-covered to an ice-free ocean would occur quite suddenly—within a few years.

We cannot yet predict what would happen following the complete melting of the arctic sea ice. One possible configuration would be that the ocean would remain permanently open except for some coastal shallow water regions, which would freeze in the winter and melt the following summer. Another possibility would be for a substantial proportion, perhaps even most, of the surface to share this freezing-melting cycle.

A critical factor in determining the relative likelihood of these two alternatives is the density structure of the resulting ocean. Near 0°C, the density of seawater is very little influenced by temperature and is largely determined by salinity. The salinity of the Arctic Ocean depends upon the relative influence of the saline, comparatively warm, Atlantic water entering from the Norwegian Sea and the balance of evaporation, precipitation, and runoff. Typically at high latitudes, precipitation and runoff exceed evaporation. With contemporary geography, where the Arctic receives the drainage from all of north-central North America and huge areas of northern Asia, this situation is likely to persist no matter what the effects of reduced ice cover in the Arctic.

The Arctic Ocean will thus continue to have relatively low salinity upper water. The depth of this upper layer will be crucial in determining the nature of the ice regime. An upper layer of only a few tens of meters might have insufficient thermal capacity to store enough heat to prevent winter freezing. However, if winds prove strong enough to mix this upper layer to a depth of appreciably more than 100 meters, no freezing would occur. It is difficult to determine, without the aid of a fairly complete model, and better knowledge of oceanic mixing processes than we now have, which situation would prevail. The Northeast Pacific, another oceanic area with low surface salinity, is mixed to a depth of over 150 meters so far as salinity is concerned, but the regime there is sufficiently different that no conclusive inferences may be drawn. (Relative to the Arctic, the Northeast Pacific has a large influx of high-salinity water and, perhaps more importantly, at over 6°C is warm enough for thermal effects to influence the density significantly.) (See the annual "Oceanographic Data Reports" from weather-ship station "P," published by the Fisheries Research Board of Canada.)

If a winter ice cover should occur, the effect on climate might be quite small with respect to the present permanently ice-covered state. If the surface remains largely ice-free all year, it is conceivable that very profound climatological changes would occur which could influence the whole globe. This influence would probably be harmful in some places and advantageous in others. Only very carefully conducted and comprehensive theoretical studies, supported by observational and experimental work designed to clarify mechanisms and establish parameters, will permit appraisal of the advantages and disadvantages of an open Arctic.

One of the most crucial problems involves the future of the Greenland ice cap. This ice mass contains so much water that to melt it would raise world sea levels by about 7 meters. Such a rise in sea level would have very serious consequences for many coastal areas of the world, and even 5 percent of it would be a major disaster for the city of Venice.

While the amounts of sea ice and snow cover on low-lying land generate a positive feedback mechanism because of their very large albedo, snow-fed glaciers at high altitude, like the Greenland ice cap, are subject to a different positive feedback mechanism. This arises because a decrease in the ice depth lowers the altitude. The surface is then immersed in somewhat warmer air, and further reduction in ice volume becomes more probable.

CONCLUSION

Positive feedback mechanisms make the regimes of arctic sea ice and the terrestrial ice caps very sensitive to climatic changes. This sensitivity makes them vulnerable to effects arising from man's activities. It is probable that a deliberate effort to remove the arctic sea ice could be successful. There is also a serious possibility that a global temperature rise of a degree or so, such as may follow man's injection of CO_2 and heat into the atmosphere within the coming century, will itself lead to melting of the arctic ice.

The consequences following from the opening of the Arctic

are difficult to appraise, but their implications for regional and global climate and for low-lying coastal areas could be profound.

RECOMMENDATION

We recommend that an international agreement be sought to prevent large-scale (directly affecting over 1 million square kilometers) experiments in persistent or long-term climate modification, until the scientific community reaches a consensus on the consequences of the modification.

It does not follow from this recommendation that smaller-scale (appreciably less than 1 million square kilometers) experiments on climate modification should not be pursued. Such experiments may furnish the best means of obtaining the empirical data necessary to provide verification of the theoretical models that must be constructed if the consensus referred to in this recommendation is ever to be reached.

7.3.2
Engineering Proposals to Modify the Ocean

From time to time proposals have been put forward which, if carried out, would directly or indirectly modify the ocean. A few of these that might have some climatological influence are discussed in this section.

ARTIFICIAL UPWELLING

As was pointed out in Chapter 4, in many parts of the ocean biological productivity is limited by lack of nutrients in the near surface water. Nutrient depletion of the upper water usually affects only the comparatively shallow summer mixed layer. Mixing to the depth of the winter mixed layer could approximately triple the supply of nutrients available to phytoplankton. Mixing upwards of the water from below, the main thermocline would yield a virtually unlimited nutrient supply. Artificial upwellings of this kind might substantially increase the productivity of the ocean.

Mechanical Mixing

Mechanical mixing of the layer of ocean water above the level of the winter thermocline (the thickness of which is variable but of the order of 150 meters) would typically in mid-latitudes lead to a decrease in later summer surface temperature by about 4°C. There would be a smaller increase in winter surface temperature (perhaps 1°C, but dependent on a complex of circumstances that would vary from place to place). Heat transport by ocean currents would change. Such mixing could be accomplished over areas small compared with the total oceanic area, but not small compared with natural upwelling regions, by using wave energy propagating in from elsewhere.

The climatic effects of creating such a small artificial upwelling would be quite large locally and would probably include the formation of a low-level summer inversion and some low-level summer fog. Both evaporation and sensible heat flux would increase in winter. Existing numerical models of the atmospheric behavior are probably adequate to elucidate the consequences of making changes in sea surface temperature of this kind.

Wave energy is insufficient to enable mixing over very large areas of the ocean, but large-scale mixing using other energy sources could perhaps become possible in the future. If such artificially mixed regions should extend over a significant fraction of the ocean, very profound climatological effects could ensue. Fortunately, however, if the climatic effects proved unfavorable, the surface modification could be eliminated and its immediate impact would probably cease within two years. It should be pointed out, though, that the sensitivity of the earth's climate might lead to effects far distant from the location of the upwelling which might be quite difficult to reverse.

Thermal Mixing

Concern for the heating of the atmosphere by artificial energy production may well lead to the use of deep ocean water as cooling water for thermal power generators (including nuclear power). It is possible to regard this deep water as the largest heat sink in the world. Deep water used in this way would be warmed, and when returned to the ocean it would seek a level nearer the surface. Suitable design would bring the water right to the surface, where it would serve the additional function of supplying the nutrients to the upper waters.

Magnitude estimates are instructive. Global energy production will soon reach 1.25×10^{13} watts. If a large fraction of this power is produced in fixed thermal generating stations, a significant proportion of it will have to be passed into some heat sink. For illustration, let us assume a quantity of 2.5×10^{12} watts of waste heat is used to heat water from 5° to 25°C. This requires 10^{18} g of water per year. The heat amounts to about 1 one-thousandth of the heat now transported meridionally by the ocean. The yearly water quantity is equal to that transported by a major ocean current like the Gulf Stream in a period of about 4 hours. The influence of such an activity thus would seem to be small. However, it should be pointed out that heat passed into the ocean in this way has not been removed from the earth. It will ultimately make its presence felt in some way in the global heat balance and cannot be subtracted from man's energy input into his environment.

TRANSPORTING ICE FROM THE ANTARCTIC

We assume that a significant fraction of the water supply of an area such as Australia might be supplied by towing rafts of antarctic ice northward. The economics of such an operation vis-à-vis other processes such as desalination are as yet unclear, but in a world becoming concerned about the effects of energy production, it may eventually prove attractive. The movements of volumes of the order of 10^{12} m^3/yr (equal to 1 meter of water on an area 1000 kilometers square, or to half the annual flow of the Mississippi River) would then be reasonable. Assuming that the displaced mass is replaced by surface water, with temperature of about 20°C, there evolves a transport of about 1.25×10^{13} watts, which is rather less than 1 percent of the natural oceanic transport. Thus large regional effects would be induced, but the global influence on the heat transport would be small. The additional irrigated area would have its effect on the albedo of the earth. The side effect thus produced would probably be of more importance than the added heat transport.

7.3.3
Surfactants

The interface between the atmosphere and seawater is the site of accumulation of surfactants produced by natural and man-influenced processes. The naturally occurring slicks are composed of high molecular weight organic species such as the fatty acids and the fatty alcohols. Oil films from the release of petroleum, 0.01 to 0.1 μ thick, form after discharges from ships or after releases from drilling operations (SCEP, 1970). It is conceivable that other natural products, as well as synthetic organic chemicals, contribute to the formation of these surface films.

These surfactants can accumulate nonpolar organic species such as the halogenated hydrocarbons, including DDT residues and the PCBs, further altering the chemical and physical properties of the air-sea interface. Convergences, always present at the sea surface, tend to alter the form of such surface films. Oil slicks are transformed into small lumps, millimeters in size, which appear to have a residence time on the sea surface of several months (SCEP, 1970) and are quite prevalent in the Sargasso Sea area. These lumps are unpleasant esthetically and may have some biological importance, but they have little or no effect on the atmosphere.

The films, on the other hand, do have some influence. Only near the source of oil are oceanic surface films of sufficient thickness to provide a barrier to evaporation or heat flow. However, they reduce the intensity both of the very short surface waves and of the turbulent motion in the top few centimeters of the water. Since the average slope of the surface is very largely determined by these short waves, slope statistics are strongly affected by the slick. (It is this slope change that makes the slick visible to the eye—the slick area gives much less diffuse backscatter of light than does a clean surface.) Albedo may be modified by a few percent. Damping of the short waves also eliminates the wave enhancement of surface transfers. This enhancement normally occurs because of the nonlinear response to the thickness of the depleted layer within the top millimeter of the water. Inhibition of near-surface turbulence has a similar effect. Experiments indicate that these surface effects result in temperature changes of about 0.5°C, and that they can have either sign. Interpretation of some data is complicated by the fact that the films occur in regions of convergence which may have differing properties for other reasons. Although it is difficult at this time to assess the effects of surface films quantitatively, the increasing number of sources of petroleum discharges and the possible importance of the chemistry of the films call for close observation.

7.4 The Land

7.4.1 Heat Released

Heat is released into the atmosphere from industrial, transportation, and domestic sources and into rivers, lakes, and oceans from industrial sources. These releases are to a large extent (perhaps two-thirds) concentrated in the densely populated urban areas that constitute only a small fraction of the land area of the world. According to the data presented in Chapter 4, the additional input of artificial (sensible) heat into the atmosphere occurs mainly on the local and regional scales where there are many heat sources producing 10 to 100 W/m^2. In areas with long cold winters, for example, Moscow or Fairbanks, Alaska, the production may be more than 200 W/m^2. In larger industrial areas such as the "Megapolis" of the eastern United States (Boston, Massachusetts, to Washington, D.C.) or the industrial area of northwestern Germany, southern Netherlands, and central Belgium, totaling an area of $4.8 \times 10^4 km^2$, the interspersed suburban, recreational, and rural areas lead to an area-average output of heat resulting from man's energy conversions of the order of 2 to 10 W/m^2. In these regions, merging of the heat-source areas seems doubtful; such a merging is apparently restricted to areas below $10^4 km^2$. In the foreseeable future, however, we must envisage a large extension of existing heat sources, the creation of major new sources, and the hitherto unaffected areas of these sources. Thus we must visualize the future development of heat sources with an additional output of 20 to 50 W/m^2 over local areas (10^4 to $10^5 km^2$).

The main effect of this heat output on the local scale is the creation of a stationary three-dimensional heat island, as described, for example, by Bach (1970) for Cincinnati. The intensity of this heat island reaches a maximum during nighttime, when the surrounding rural areas are cooling under the effect of the net outgoing (terrestrial) radiation. The minimum intensity occurs shortly after noon, when the sensible heat flux of the surrounding areas more or less equals the input of artificial heat. Under stable conditions with light winds, the effect is restricted to a shallow atmospheric layer a few hundred meters in thickness, the surface warming being about 2° to 6°C.

According to V. Bjerknes's circulation theorem, any local heat source of sufficient intensity produces a thermally induced circulation that can be directly observed when the synoptic-scale winds are very weak. In contrast to the usual type of thermally induced circulations, such as land and sea breezes, this thermal circulation cannot reverse its direction but has a peak of intensity during nighttime and a minimum intensity in the afternoon. The intensity of this country breeze is very weak, usually 1 to 2 m/sec. Assuming a diameter for a circular heat source of 10 (100) km, we obtain with a surface wind of 1 m/sec a convergence of 2.10^4 (2.10^5) sec^{-1}. If these conditions occur in a surface layer 500 m in thickness, an average vertical velocity above the heat island of 10 (1) cm/sec is produced. This causes a local destabilization of the surface layer. In some cities where there may also be an increase in water vapor output from cooling towers, hot water channels, and so on, there may be an increase of frequency of showers and thunderstorms above or downwind from the heat sources. An evaluation of the radar echo frequencies in a 100–km circle around Bonn, using 503 radar-screen photographs from 17 days with widespread shower activity, revealed a local increase in the shower frequency at the southern downwind border of the industrial region of western Ruhrgebiet (Claasen, 1970). Detailed maps of *average* rainfall also show this effect (for example, the Ruhrgebiet or over Washington, D.C.), although this is not a universal result; in some cities, the local moisture sources are reduced by replacement of forest by concrete. Finally, the observed increase in the frequency of weak rainfall or drizzle over industrial areas under thermodynamically stable synoptic conditions may be due to the high particle content and partly to the increase of precipitable water by artificial evaporation.

The effects of urban and regional heat sources on continental and global climate are uncertain. Cold outbreaks in late autumn and winter over the Great Lakes, the Gulf Stream, and the Sea of Japan have a significant effect on large-scale weather processes, distorting the tropospheric flow patterns and creating troughs of low pressure. In these cases, however, the sensible and latent heat fluxes from the open water are of the order of 500 W/m^2. In comparison, the megalopolis man-made heat-source strength, as mentioned previously, amounts to only about 2.5 to 10 W/m^2; it is likely to increase somewhat in the next century, and its areal extent will certainly expand. We can only speculate, therefore, that there is a *possibility* that urban complexes may soon, or even now, occasionally trigger changes in the large-scale weather patterns. We consider, however, that when computing average trends over decades these effects would be largely obscured by the presence of natural fluctuations.

Some clues about the solution to this problem may be obtained by modeling, in which a numerical experiment is undertaken twice, once with the city heat sources ignored, as a control, and once with them included. A preliminary experiment along these lines has been reported by Washington (1971). He introduced an additional surface heat flux of 25 W/m^2 over all the land areas of the world, a figure that is unrealistically large, but he comments that "we would prefer extreme experiments to see at least some quantitative effect." The artificial heat source was introduced at day 35, and the experiment was continued to day 57. The root-mean-square difference in surface temperature between the two cases increased rapidly at first, but by day 50 it began to level off at about 5°C. Zonally, temperatures in the tropical regions were not greatly affected, the maximum departures being near the North Pole. This kind of research should be encouraged, and in the second-generation models the effects of the zonality of population and of industrial distributions should be included.

7.4.2
Ground Cover

URBANIZATION

The effects of man-made heat sources and of snow removal have been considered in previous sections. Additional urban climatic changes are caused by the increased surface roughness, the changed albedo, the accelerated runoff, and the changed heat-storage capacity resulting from the replacement of forest and fields by concrete and buildings. These changes are well illustrated in Table 7.2. Although the microclimate has been greatly modified from

Table 7.2
Average Changes in Climatic Elements Caused by Urbanization

Element	Comparison with Rural Environment
Contaminants:	
condensation nuclei and particles	10 times more
gaseous admixtures	5 to 25 times more
Cloudiness:	
cover	5 to 10 percent more
fog—winter	100 percent more
fog—summer	30 percent more
Precipitation:	
totals	5 to 10 percent more
days with less than 5 mm	10 percent more
snowfall	5 percent more
Relative humidity:	
winter	2 percent less
summer	8 percent less
Radiation:	
global	15 to 20 percent less
ultraviolet—winter	30 percent less
ultraviolet—summer	5 percent less
sunshine duration	5 to 15 percent less
Temperature:	
annual mean	0.5° to 1.0° C more
winter minima (average)	1° to 2° C more
heating degree days	10 percent less
Wind speed:	
annual mean	20 to 30 percent less
extreme gusts	10 to 20 percent less
calms	5 to 20 percent more

Source: Landsberg, 1970.

its natural state, we do not believe that the influence of these latter processes will extend significantly beyond the built-up areas.

ROADS

Relative to the terrain they replace, roads are usually dry, thus affecting the heat budget. Also, gravel and concrete road surfaces will generally have a much larger albedo than their surroundings, while asphalt surfaces may produce comparable or somewhat lower values. In moist areas, a dry road can provide a heat source that will have an associated convective plume and serve as a daytime trigger for local cumulus cloud formation and perhaps also for precipitation. These effects could be studied in regional models of climate, simulating the roads by line sources of heat.

VEGETATION

Modification of the natural vegetation affects several significant climatic parameters: the surface roughness, the surface albedo, and the apportionment of the available net radiation into sensible and latent heating of the atmosphere. Because tree roots usually penetrate rather deeply into the soil, they are frequently not under moisture stress at times when grasslands have dried up. In fact, there are cases in which the cutting of a forest has caused the water table to rise to such an extent that the area has turned to swamp. The suggestion has been made facetiously that a way to conserve water on the east slopes of the Rocky Mountains would be to replace the forest cover by artificial Christmas trees, thus eliminating transpiration but permitting the spring snow melt to continue to occur slowly. Attempts have been made to spray forests with antitranspirants: although some chemicals have proved effective in greenhouse experiments, aircraft spraying has not been particularly successful, largely because of the difficulty in delivering the spray to a significant fraction of the leaf cover in a forest.

The heat balance of a vegetative volume is given by the equation:

$$Q_S = F_A + F_L + F_G + F_S$$

where Q_S is the net radiation, F_A is the convective heat transfer at the top of the canopy, F_L is the latent heat transfer at the top of the canopy, ($F_L = L \cdot ET$, where L is the latent heat of evaporation and ET is the evapotranspiration), F_G is the conductive heat transfer at the soil surface, and F_S is the rate of storage of heat in the vegetative cover. For climatic averages, the quantities F_G and F_S are negligibly small, the surface coming rather quickly into equilibrium following after air-mass changes.

The magnitude of Q_S depends not only on conditions in the free atmosphere but also on the value of the surface albedo. However, the replacement of a forest by grassland or semidesert has only a rather small effect on the albedo (excluding the special effects of snow), as indicated in Table 7.3. Maps of the surface albedo pattern for North America have been published by Kung and his associates (1966).

The apportionment of net radiation into sensible and latent heat is usually discussed in terms of the Bowen ratio (F_A/F_L), the value of which usually ranges from 0 (entirely latent heat transfer) to infinity (entirely sensible heat transfer), although negative values can occur when the latent heat term exceeds the available net radiation and must be partly supplied by cooling the air. In humid regions, the Bowen ratio varies between 0 and 1, and under very humid conditions—for example, above irrigated rice fields, swamps, and above tropical rain forests—the value fluctuates around 0.1, as it does also above tropical oceans. Under arid conditions, the Bowen ratio lies well above 1, averaging about 10 in desert regions. In an extreme desert without traces of precipita-

Table 7.3
Climatological Estimates of Albedo over Land

Nature of Surface	Albedo
Steppe and coniferous forests in months when mean temperature > 10° C	0.13
Deciduous forests in months when mean temperature > 10° C	0.18
Tropical forests during dry months	0.18
Savannahs and semideserts during dry months	0.14
Tropical forests during wet months	0.24
Savannahs and semideserts during wet months	0.18
Deserts	0.28

Source: Budyko, 1971.

tion and without groundwater, the value approaches infinity. Monthly world maps of F_A and F_L are available in the well-known heat budget atlas (Budyko, 1963). Figure 7.5 shows a map of the annual distribution of the Bowen ratio over the North American continent. Similar maps of all continents are now under construction at the University of Bonn by D. Henning.

Table 7.4 illustrates the changes that take place after conversion of a temperate-zone forest to agricultural fields, assuming incoming shortwave radiation of 125 W/m² and an effective outgoing infrared radiation of 50 W/m² over the fields (Mitchell, 1970). The differences are certainly significant locally. Taking an average over all continents, the annual reduction in transpiration due to new deforestation is probably of about the same order of magnitude as the increase in evaporation due to new irrigation. However, these two man-made changes are occurring in different parts of the world.

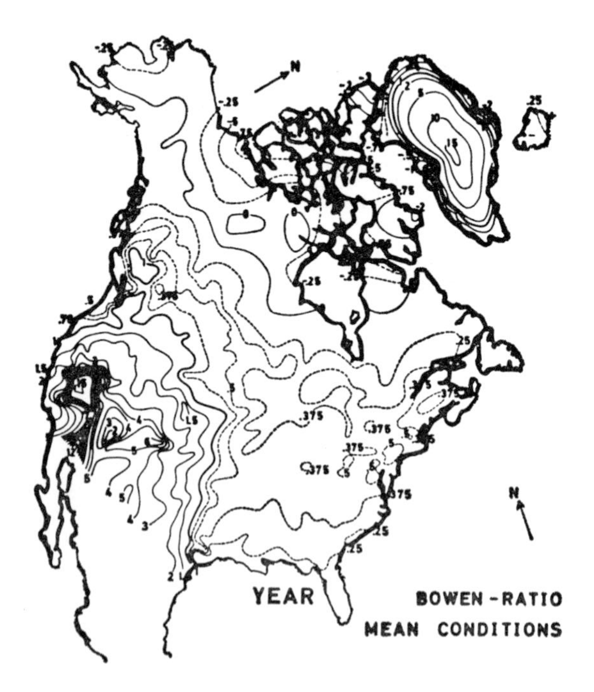

Figure 7.5 Annual distribution map of the Bowen ratio over North America. Source: Budyko, 1963.

Table 7.4
Changes of Heat Budget after Conversion from Forest to Agricultural Use

	Albedo Assumed as Representative	Bowen Ratio	Q_S (W/m²)	F_A (W/m²)	F_L (W/m²)	ET (mm/month)
Coniferous forest	0.12	0.50	60	20	40	41
Deciduous forest	0.18	0.33	53	13	39	40
Arable land, wet	0.20	0.19	50	8	42	43
Arable land, dry	0.20	0.41	50	15	35	36
Grassland	0.20	0.67	50	20	30	31

Note: At the present time, little replacement of temperate forests by arable lands is in progress. The process may now be largely complete, and indeed in some areas is being reversed as marginal farmland is returned to forest. However, in some tropical regions deforestation is proceeding rapidly, resulting in some cases in changes of Bowen ratio even more dramatic than in the temperate zone.
Sources: Mitchell, 1970; Flohn, 1971 (unpublished).

In particular, the average annual value of the Bowen ratio over tropical jungles is about 0.3 (Budyko, 1971) whereas it ranges from 2.0 to 6.0 over the Rajputana Desert of western India, a region thought to be covered formerly by vegetation (see discussion later). Destruction of the Brazilian, Indonesian, or the African jungles might be expected to produce regional increases in Bowen ratio of almost the same order of magnitude. The replacement of such large latent heat sources by large sensible heat sources might have a significant effect on the generation and dissipation of tropical easterly waves. This deforestation might also, as suggested by Newell (1971), affect the dynamics of the general circulation through a series of nonlinear interactions. We believe, therefore, that there is an urgent need for numerical experiments designed to test the sensitivities of the models to major changes in the reapportionment of sensible and latent heat fluxes in the tropics.

We should mention here that the destruction of a jungle is almost an irreversible process.* Soil moisture is no longer retained,

* This is true also in the subarctic. The microclimate is so greatly affected (increased wind speed and a much shorter frost-free period) that the forest returns to its original position only with the greatest of difficulty. As a further hindrance, browsing animals eat the young seedlings at the edge of the forest.

and there are frequent flash floods during the rainy season. In addition, the soil dries rapidly and may release clouds of dust into the atmosphere, changing the regional radiation balance. Bryson (1971), for example, has argued that the Rajputana Desert in western India is man-made. Noting that the vertically integrated water vapor content being comparable with that over some tropical forest regions, he suggests that the tropospheric dust loading has increased atmospheric subsidence, thus inhibiting precipitation. Archaeological and pollen studies indicate that the desert was indeed relatively fertile several thousand years ago and that it contained a freshwater lake that supported an early civilization. Bryson's studies remind us that numerical simulations of man-made changes in the tropics must include the effects on the tropospheric radiation balance induced by windblown dust.

Finally, mention should be made of the fact that the value of the Bowen ratio is decreased by irrigation. The areas now under irrigation are estimated to be about 2×10^6 km² (Budyko, 1971), or about 1½ percent of the total area of the continents. Only some of these irrigation projects are situated in arid climates, but there the most dramatic changes in the heat budget are occurring. Table 7.5 illustrates the man-made variation of the heat budget in the Sahara oases of southern Tunisia (Flohn, 1970), where a negative value of the Bowen ratio occurs (the latent heat flux exceeds the available net radiation). Because the climatic causes of deserts are large-scale atmospheric subsidence and the rareness of rain-producing synoptic events, the increase of atmospheric humidity above irrigated areas can hardly result in increased precipitation.

Table 7.5
Annual Heat and Water Budget of Tunisian Oases

	Area (km²)	Estimated Albedo	Bowen Ratio	Q_S (W/m²)	F_A (W/m²)	F_L (W/m²)	ET (cm/yr)	Precipitation (cm/yr)
Oases (average)	150	0.15	−0.26	100	−36	136	168	15
Semi-desert	35,000	0.20	+5.6	80	+67	12	15	15

Source: Flohn, 1970.

7.4.3
Surface Water Area

Man is changing the total area and distribution of surface waters by construction of artificial lakes and reservoirs, by draining swamps, by river "training" (that is, engineering work such as the stabilization of banks), by flood control, and by evaporation of seawater to produce salt.

ARTIFICIAL LAKES AND RESERVOIRS

At present there are about 300 artificial lakes and reservoirs having surface areas of more than 100 km^2 each. Future projections are uncertain, but there is no doubt that the number of artificial lakes will increase in response to the needs of industry, power generation, and agriculture.

Some of the possible effects of artificial lakes and reservoirs on climate are similar to those arising in oases and irrigated areas. These effects have already been discussed in the previous section. We should note here, however, that on the basis of recent estimates (Fels and Keller, 1971), the total water surface area of man-made lakes and reservoirs may be of the order of 300,000 km^2. Assuming the man-caused annual surplus evaporation (as compared to the natural evaporation of the reservoir sites) is of the order of several hundred millimeters, the latent heat annually required is of the order of 10^{17} calories, which is some 10 percent of the corresponding value characterizing the effect of irrigation. It should be mentioned, however, that artificial lakes and reservoirs are distributed over the whole globe and are quite separate from one another. Therefore, their effects are also separated, and thus can be considered on a local scale. Nevertheless, we feel that a census of artificial lakes should be prepared regularly, in order that trends may be examined.

DRAINAGE

The drainage of swamps affects the water, radiation, and heat balances of the earth's surface in the sense opposite to that of artificial lakes. Because drainage activities occur predominantly in humid areas where conditions suitable for rapid evaporation are infrequent, their effects are less significant than those of man-made lakes. In populated areas, the drainage of swamps is largely complete. Because of the increasing population of the earth, more swamps will be drained and converted to agricultural land but the affected area will not be too large. About 80 percent of the remaining undisturbed land bank is covered with forest, which may be easier than swamps to change into cropland (see SCEP, 1970, Tables 2.1 and 2.2),

RIVER ENGINEERING AND FLOOD CONTROL

River engineering is undertaken mostly in populated areas. By eliminating seasonal floods, relatively large amounts of land have become available for agricultural and urban use. The net effect on climate, however, is not significant except on the microscale.

SALT PRODUCTION

The evaporation of seawater to produce salt commercially takes place mainly in arid seacoast areas. Because of geographical constraints, we can assume that even considering the rapid increase of the population of the world, and the need for more salt, there will not be a sharp increase of the area used for salt production.

WATER CONSUMPTION

The increasing water used by man for domestic and industrial purposes and the even larger use for irrigation leads to an acceleration of the hydrological cycle over land (see Section 5.4). The intensity of this process depends on two quantities: the additional nonreturnable water consumption of man per head V_a and the population density P. If V_a is given in liters per day per person and P in persons/km^2, the additional evaporation E_a is given by

$$E_a = V_a \cdot P \; (10^{-7}\ \mathrm{cm/day})$$

Keller (1970) recently reevaluated the water balance of the Federal Republic of Germany, comparing the period 1931 to 1960 with the period 1891 to 1930, and obtained an increase of precipitation of 3 percent, a decrease of runoff of 12 percent, and, as a residual, an increase of evapotransportation of about 20 percent, or 8 cm/yr. This has been attributed to a rise of agricultural production as well as of industrial and domestic water demand.

Using Lvovich's (1969) figure for nonreturnable water consumption of 1800 km^3/yr for 1965, we obtain a value for V_a of about 1500 liters/day as a global average, of which more than 90 percent is used for irrigation. Using the preceding formula for western and central Europe ($1.56 \cdot 10^6$ km², $P = 169/\text{km}^2$), we obtain $E_a = 9$ cm/yr, nearly coinciding with Keller's result.

Assuming as annual increase of water consumption of 3 percent (German 2.6 percent), we obtain for the year 2000 A.D. a value of about 4900 km³/yr (Lvovich, 1969). Assuming a predicted world population of 6×10^9, this would yield $V_a = 2240$ liters/day of which about 75 percent will be used for irrigation. For western and central Europe, we may assume a more conservative rise of population density of only 50 percent, from 1965 to 2000 A.D.: this yields $E_a \sim 21$ cm/day, or nearly half of the natural evaporation. If the assumed rates are slower, the result may be postponed by 10 or 20 years: a substantial acceleration of the hydrological cycle certainly will take place, however, with a weak but increasing effect also on the oceans. The amount of additional evaporation, nevertheless, is limited by the available net energy, so the increase envisaged by Lvovich cannot be extrapolated beyond 2000 A.D.

7.4.4
Runoff Control and Water Management

Man affects the hydrologic cycle of watersheds and river systems. Some of these surface changes are undertaken with the direct intention of controlling the runoff (man-made lakes, watershed management, artificial recharge of groundwater); others alter the runoff indirectly (urbanization, soil cultivation, changes in vegetation cover).

Results of a global appraisal indicate that about one-third of the 37,000 km³ total annual runoff from the land areas is regulated by the natural storage effects of river basins while a further 10 to 12 percent is controlled by man-made lakes and other human interventions. Recent estimates on global water uses are summarized in Table 7.6 (Lvovich, 1969). As the differences between the withdrawals and returns indicate, irrigation is by far the most significant consumption, causing a 5 percent decrease in the natural runoff and 215 percent increase in evaporation from land areas. The estimated large increases and the redistribution of the four principal water uses by the year 2000 indicate that serious consideration should be given to the climatic implications of this question.

Water management and runoff control may considerably change the natural water balance even in relatively large areas.

Table 7.6
Estimated Global Water Needs and Water Resources (km^3/yr)

	1965		2000	
	With-drawal	Return	With-drawal	Return
Needs:				
Municipal water supply	98	56	950	760
Irrigation	2300	600	4250	400
Industry	200	160	3000	2400
Energy	250	235	4500	4230
	2848	1051	12,700	7790
Resources:				
Precipitation on continents	108,000			
Runoff	37,000			
Evapotranspiration	71,000			
Evapotranspiration from agricultural areas	(3560)			

Source: Lvovich, 1969.

According to recent estimates, the runoff from the Federal Republic of Germany, for example, has decreased by 12 percent between the periods 1891 to 1930 and 1931 to 1960, primarily due to increased water uses (Keller, 1970). In most climates, these changes will have little effect on regional weather; in arid zones no substantial surface-layer change is to be expected. The consequences for global climate require careful evaluation, however, using models capable of investigating the effects of trends not only in the hydrologic but also in the other geophysical inputs.

7.4.5 Mining of Groundwaters

The subsurface reservoirs of the earth were filled up gradually during geological history, and there is evidence that this process is now being reversed by extraction rates of water surpassing the rates of natural recharge. Indications of simultaneous increases in the amounts of water in the world ocean and the polar ice masses observed during the last 80 years may be attributed—at least partially—to the groundwater extractions that may in turn lead to a considerable decrease in the groundwater table in many parts of the world (see Section 5.4).

7.4.6
Conclusion

Many of man's activities affect the heat budget of the atmosphere and thus can result in inadvertent modification of the climate. In most cases, at least at present, effects are predominantly local and regional, but there are some global effects. Of obvious importance are those activities that directly influence the amount of heat available to drive atmospheric motions, in particular, the production of heat by energy generation and the absorption of solar energy by changes in the surface albedo.

Less easy to appraise are those numerous activities, including such things as irrigation and deforestation, which change the apportionment of heat release from the surface between sensible heat and latent heat of vaporization (that is, the Bowen ratio). Sensible heat is immediately available to drive atmospheric motions. The latent heat becomes available ultimately only when the vapor condenses and precipitates. This usually occurs at an appreciably different altitude and longitude from the place of evaporation, and frequently at a different latitude. Comparatively large amounts of heat are involved in these transfers and their importance should be studied.

7.4.7
Recommendation

We recommend that model studies be undertaken to reveal the consequences of altering the ratio of direct heating of the surface and of evaporation (the Bowen ratio), for example, in marginal semiarid regions.

7.5
Climatological Census*

7.5.1
Discussion

We feel that the study of climate and of man's impact on climate will receive increasing emphasis in the next few years. These studies will require global data in as homogeneous a form as possible,

* The word "census" is used here not in its narrow sense of population count but in the broader sense used by contemporary governments: a periodic collection of data of many kinds such as are required to define the status of a nation —or in this case the earth's climate—at a particular time.

Table 7.7
Data Needed for a Climatological Census

Factor	Frequency of Census	Space Average
Factors describing the state of the climate:		
Arctic ice cover (When a suitable technique is developed, should include thickness)	Yearly (August)	Arctic Ocean
Mass of glaciers	10 years	Hemisphere
Sea level	10 years	Global
Groundwater volume	10 years	Continents
Biomass of trees	10 years	Continents
Natural freshwater bodies (area and volume)	10 years	Continents
Volcanoes (now being collected)	Yearly	Latitudinal zones
Factors describing man's impact:		
Irrigation area	Yearly	Continents
Artificial lakes (area and volume)	5 years	Continents
Urban area	5 years	Continents
Fuel consumption	Yearly	Continents
Forest fires	Yearly	Continents
Factors necessary for control of experiments:		
Cloud seeding event inventory	Monthly	Continents

and it would greatly facilitate this necessary work if the required data were collected and organized in a systematic way.

We do not employ the word "monitoring" to describe all the activities we envisage because "monitoring" carries a connotation of "continuously watching over" which is unnecessary for some factors and impractical for others. Monitoring in this narrow sense is indeed a reasonable approach for many factors. However, what is needed for others is a periodic appraisal of the status of the factor. The frequency with which information should be gathered and the area of over which averaging should be performed will vary according to the factor.

Some of the required data can be collected by satellite observation. In general, where satellites can be used they should be, even if the costs are somewhat higher. (The advantages of remote sensing of this kind are not just, or even chiefly, that frequent and rapid surveys may be made but that the data are essentially homogeneous, no matter from where collected.)

We can identify at least three types of data needed for climate studies: data relevant to a description of the existing state of the earth's climate, data relevant to the nature and pace of man's activities that affect climate, and data needed to ensure the proper control of experimental and observational programs related to climatic change.

7.5.2 Recommendation

We recommend that international uniformity be achieved for data relevant to the study of climate change and that these data be made readily accessible to scientists of all countries.

Wherever possible, not only current but also historical information should be collected. In particular, climatological census data of at least the factors given in Table 7.7 should be included.

7.6 General Conclusions

We have considered those aspects of the interaction between the atmosphere and the surface of the planet which play a part in controlling the climate in the context of both natural climatic changes and of the possibility of intentional and inadvertent interferences by man.

The most sensitive surface state appears to us to be the ice and snow, particularly arctic sea ice, because of the large change of albedo accompanying a change in its area and the relative ease of modification. Inadvertent change in the next four decades will be mainly of regional importance. However, over the only somewhat larger time scale of the next century we recognize a real possibility that a global temperature increase produced by man's injection of heat and CO_2 into his environment may lead to a dramatic reduction or even elimination in arctic sea ice. Further, there is a possibility that deliberate measures to induce arctic sea ice melting might prove successful and might prove difficult to reverse should they have undesirable side effects.

The example of the arctic sea ice is an interesting illustration of the sensitivity of a complex and perhaps unstable system that man might significantly alter over the next few decades by making relatively small modifications in the earth's present heat balance. Some models indicate that a few degrees change in average temperature of the Northern Hemisphere might begin a melting of the arctic sea ice.

Some studies indicate that large areas of open arctic sea would tend to cause melting of the remaining sea ice, with the eventual result that it would disappear entirely. Once gone, it probably would not readily freeze over again. The melting of the arctic sea ice would not affect ocean levels. The changes in climate, especially in the Northern Hemisphere, which would occur after the arctic sea ice melted are unknown, but they might be large and include changes in precipitation, seasonal temperatures, wind systems, and ocean currents. The possible effects on the Greenland ice cap are likewise unknown: whether it would increase because of an increase in precipitation or begin to melt, and how long it might take for such changes to occur, although we believe any such changes would take many centuries. A beginning has been made in studying the sensitivity of arctic sea ice to temperature changes by the use of models.

With respect to man's impact on land surfaces, we have considered possible effects of the heat release resulting from all man's activities in energy conversion: "useful" as well as "waste" energy eventually finds itself heating the environment. We see the possibility of climatically significant changes on a regional scale in the near future. On the local scale, this influence is already very large. Local and regional scale effects are amenable to study by existing climatic models.

Finally, we have considered man's manipulation of groundwater and his use of surface waters. These produce large local climatic changes, for example, when an irrigated region becomes cooler. On a global scale, the effects are principally upon albedo and on altering the amount of evaporation that takes place. We suggest that these effects could be studied with the use of existing models, but it seems that most of the work remains to be done. In particular the climatic impact of deforestation of tropical jungle, which seems to result in sharply reduced evaporation and increased local heating, requires study. We note that "mining" of fossil groundwater is leading to loss of a nonrenewable resource, and we note that this and inadvertent tapping of other groundwater supplies may be depleting them to the extent that they may be causing the recent rise in sea level.

References

Arnold, K., 1961. An investigation into methods of accelerating the melting of ice and snow by artificial dusting. *Geology of the Arctic,* edited by G. O. Raagh (Toronto: University of Toronto Press), pp. 989–1012.

Bach, W., 1970. An urban circulation model, *Archiv für Meteorologie, Geophysik und Bioklimatolagie, Serie B, 18:* 155–168.

Bryson, R. A., 1966. Airmasses, streamlines and the boreal forest, Technical Report No. 24, Department of Meteorology, University of Wisconsin, Madison.

Bryson, R. A., 1971. Climatic modification by air pollution, Preprint, Conference on Environment Future, Helsinki, Finland, 36 pp.

Budyko M. I., 1963. *The Heat Budget of the Earth* (Leningrad: Hydrological Publishing House).

Budyko, M. I., 1971. *Climate and Life* (Leningrad: Hydrological Publishing House).

Clarke, J. F., 1969. Nocturnal urban boundary layer over Cincinnati, Ohio, *Monthly Weather Review, 97:* 582–589.

Claason, Christa, 1970. Untersuchungen über die Häufigkeitsverstärkung von Niederschlagsechos im 100 km-Umkreis um Bonn mittels 3 cm-Radar, Diplomarbeit, University of Bonn (unpublished).

Donn, W., and Shaw, D., 1966. The heat budgets of an ice-free and ice-covered Arctic Ocean, *Journal of Geophysical Research, 71:* 1087–1093.

Fels, E., and Keller, R. 1971. World register on man-made lakes, COWAR Symposium on Man-Made Lakes, Knoxville, Tennessee (unpublished).

Flohn, H., 1964. Grundfragen der Paläoklimatologie im Lichte einer theoretischen Klimatologie, *Geologische Rundschau, 54:* 504–515.

Flohn, H., 1970. Etude des conditions climatiques de l'avance du désert. (To be published in *World Meteorological Organization Technical Note.*)

Flohn, H., and Korff, H. C., 1969. Zusammenhang zwischen dem Temperaturgefälle Äquator-Pol und den planetarischen Luftdruckgürteln, *Annalen der Meteorologie,* N.F. 4: 163–164.

Hare, F. K., 1951. Some climatological problems of the Arctic and sub-Arctic, *Compendium of Meteorology* (Boston: American Meteorological Society), pp. 952–964.

Keller, R., 1970. Symposium of World Water Balance, International Association of Scientific Hydrology, *Publications, 93:* 300–314.

Landsberg, H. E., 1970. Climates and urban planning, *Urban Climates* (Geneva: Secretariat of the World Meteorological Organization), p. 372.

Lvovich, M. I., 1969. *Vodnie Resoursi Buduschevo* (Water Resources of the Future), Prosreschenie ed., Moscow. In Russian.

Mitchell, J. M., Jr., 1970. A preliminary evaluation of atmospheric pollution as a cause of the global temperature fluctuation of the past century, *Global Effects of Environmental Pollution,* edited by S. F. Singer (Dordrecht, Holland: Reidel Publishing Company; New York: Springer-Verlag), pp. 139–155.

Munn, R. E., and Bolin, B., 1971. Global air pollution—meteorological aspects, *Atmospheric Environment, 5:* 363.

Newell, R. E., 1971. The Amazon forest and atmospheric general circulation, *Man's Impact on the Climate,* edited by W. H. Matthews, W. W. Kellogg, and G. D. Robinson (Cambridge, Massachusetts: The M.I.T. Press), pp. 457–459.

Oke, T. R., and East, C., 1971. The urban boundary layer in Montreal, *Boundary-Layer Meteorology, 1:* 411.

Pasquill, F., 1962. *Atmospheric Diffusion* (London: Van Norstrand Company, Ltd.).

Peterson, J. T., 1969. The climate of cities: a survey of recent literature (Washington, D.C.: National Air Pollution Control Administration, U.S. Government Printing Office).

Ryd, 1970. Building climatology, *World Meteorological Organization Technical Note 109.*

Study of Critical Environmental Problems, 1970. *Man's Impact on the Global Environment* (Cambridge, Massachusetts: The M.I.T. Press).

Stern, A. C. (ed.), 1971. *Proceedings of the Symposium on Multiple Source Urban Diffusion Models* (Chapel Hill, North Carolina: University of North Carolina).

Washington, W. M., 1971. On the possible uses of global atmospheric models for the study of air and thermal pollution, *Man's Impact on the Climate,* edited by W. H. Matthews, W. W. Kellogg, and G. D. Robinson. (Cambridge, Massachusetts: The M.I.T. Press), pp. 265–276.

8 Modification of the Troposphere

8.1 The Planetary Radiation Field and Global Albedo

In this chapter we shall consider in some detail the processes by which man might influence climate through those activities that affect the troposphere.

If the net input of solar radiation is reduced, the planet will become cooler; if it is increased, the planet will become hotter. The climate will change, either by an overall change in mean temperature of the atmosphere and waters or (and this is almost certainly true) by a change in the circulations of the atmosphere and oceans which redistribute the solar heat before it is radiated back to space. If those properties of the atmosphere and ocean which determine the radiation to space could be changed without changing the net solar input, the redistribution of solar heat and the planetary temperature distribution would be changed, although the mean temperature of the planet would remain the same. Therefore, we look for ways in which man might be influencing the interconnected processes that determine the net solar input, both in sum and in distribution, and the distribution of the balancing outgoing radiation. If he is influencing these processes, he is changing the climate—how significantly, by how much, and in what way, these are the questions that need to be investigated.

Since man cannot change the output of the sun, he can change the net heat input only by changing the reflectivity of the planet—the global albedo. It is clear that an efficient method of changing the global albedo is to change the areas of the planet that have a high reflectivity: the snow-, ice-, and cloudcovered regions. We have considered possible changes in snow and ice cover in Chapters 6 and 7.

In this chapter we look for processes that might change the extent of cloud cover or the reflectivity of clouds. We identify possibilities of climate impact associated with the emission of particles into the atmosphere from surface sources and the emission of particles and water vapor by high-flying aircraft. We also find that man's production of particles might significantly increase the global albedo, and we identify certain areas of research required to confirm this. For this reason we find it necessary to discuss at some length the nature, amount, and radiative properties of man-made particles.

In addition to modification of the global albedo, particles also may modify the atmosphere's internal radiative processes. Small changes in the concentrations of some minor gaseous constituents of the atmosphere can also modify these processes. Carbon dioxide is the best known and the most climatically significant of these gases. It is certain that man is increasing the CO_2 content of the atmosphere, and that the initial climatic impulse of an increase of CO_2 is to increase the global average of temperature near the surface.

8.2
Particles in the Atmosphere

The atmosphere is observed to contain suspended particles. This is true wherever measurements have been made, both near the earth's surface and at all levels of the troposphere and stratosphere. The radius of these particles ranges from 10^{-7} cm to 10^{-2} cm; and they are known to be produced directly by man's activities, by natural processes in no way connected with man's presence on earth, and by natural processes which could be, and in some cases clearly have been, triggered or intensified by man. In some of the formation processes the particles are formed and released at the land or ocean surface, in others the formation is through chemical reactions among gases in the free atmosphere.

As we shall see when considering in detail the emission and formation of particles (Section 8.3), it is not possible to draw a firm distinction between man-made and natural particles, but it is useful for the purpose of this report to be able to speak of "clean air" and "polluted air." Using measurements of particle number made in remote areas (the polar regions, the central ocean areas, the high troposphere) as reference points, we suggest an arbitrary definition of "clean air" to be air with a total particle content, measured by condensation nucleus counter, of less than 700 cm^{-3}; other air is "polluted."

With this classification polluted air is considered to be the result of natural pollution and of direct or indirect man-made pollution. Natural pollution is created by processes over which man has no control. These pollutants include volcanic dust; sea spray and its particulate products; wind-raised dust from sparsely populated arid areas; particles of organic material formed from the exudations of vegetation such as terpenes in areas of coniferous forest; living material such as microorganisms, pollen, spores; and smoke from "natural" forest fires.

Direct man-made pollution is pollution by processes in which the atmosphere is deliberately used by man for disposal of waste products. Major pollutants include particles directly emitted during combustion and particles formed in the atmosphere from gases emitted during combustion. These are the products of industrial and domestic consumption of fuel, internal combustion engines, incineration of domestic waste and agricultural burning, particularly "slash and burn" practices. Some mineral particles are emitted during some industrial and mining processes. Indirect man-made pollution is a by-product of man's activities. An example is the dust raised as a result of food production practices such as ploughing and overgrazing of arid and semiarid lands. Present estimates indicate that between 5 and 45 percent of all the particulate matter in the atmosphere is produced by man (Table 8.1).

The existence of particles in the atmosphere is a dynamic process. After emission or formation they may interact with one another by collision and coagulation (the latter is a particularly important process for the smaller particles of the size range), and they are removed by diffusion to surfaces, impaction on surfaces, and in the processes that accompany rainfall and snowfall. The term "rainout" is used for processes that incorporate atmospheric pollutants in growing cloud particles that are subsequently removed by precipitation. The word "washout" is used for the removal of atmospheric pollutants that are swept out by falling rain or snow. Since a particle may lose its identity without being removed from the atmosphere, it is not a simple matter to specify the lifetime of atmospheric particles, particularly at the smaller end of the size range, but matter that enters the atmosphere in particulate form or is transformed into particles in the atmosphere may remain there for a time ranging from some minutes to days near the surface, or to years in the stratosphere.

The particulate load in the atmosphere reacts on the climate in two different ways. First, particles change the radiation field by scattering sunlight, some of it back to space, or by absorbing sunlight. The sunlight absorbed and backscattered does not reach the surface and cannot be absorbed and or rescattered there, as it would be in the absence of particles. The particles may thus change the total sunlight scattered by the earth to space—the global albedo—but the sign of the change is not obvious as it depends on the optical properties of both the particles and the underlying surface. The particles also radiate energy in the infrared spectrum, and modify the field of terrestrial radiation in a manner that again is not obvious but depends on the properties of the particles and the temperature structure of the surrounding atmosphere. The radiative effect of the particles reacts on this temperature structure and may significantly affect the static stability of the atmosphere.

Particles also affect the processes of condensation in the atmosphere that result in the formation of cloud, snow, and rain, because in the atmosphere water vapor condenses, without exception, on a particle. For a given quantity of available water vapor, the number and nature of cloud droplets formed depend on the number of the particles available for condensation. The likelihood and perhaps the amount of precipitation, as well as the radiative properties of the clouds, depend on the number and nature of the droplets they contain.

8.3
Sources of Particles

8.3.1
Emissions in the Surface Layer

Estimates of the magnitude of the emission or formation of particles less than about 20-μ radius in the surface layer of the atmosphere are given in Table 8.1 in terms of the range of values to be found in the literature. (Only one estimate was appropriate for sea salt, as will be explained later.) The wide range in many of these estimates—over an order of magnitude in the instance of

Table 8.1
Estimates of Particles Smaller than 20-μ Radius Emitted into or Formed in the Atmosphere (10^6 metric tons/year)

Natural	
Soil and rock debris[a]	100–500
Forest fires and slash-burning debris[a]	3–150
Sea salt	(300)
Volcanic debris	25–150
Particles formed from gaseous emissions:	
Sulfate from H_2S	130–200
Ammonium salts from NH_3	80–270
Nitrate from NO_x	60–430
Hydrocarbons from plant exudations	75–200
Subtotal	773–2200
Man-made	
Particles (direct emissions)	10–90
Particles formed from gaseous emissions:	
Sulfate from SO_2	130–200
Nitrate from NO_x	30–35
Hydrocarbons	15–90
Subtotal	185–415
Total	958–2615

Sources: Discussed at length in Section 8.3.1.
[a] Includes unknown amounts of indirect man-made contributions.

forest fires and slash-burning debris—is indicative of the paucity of precise data on global emissions and of the disparity in the assumptions made to arrive at the estimates.

In assessing the magnitude of the entry of soil and rock debris into the atmosphere through natural weathering processes, Goldberg (1971) extrapolated the rates of accumulation of windborne materials in glaciers and in deep-sea sediments to obtain a global average and arrived at a range of 100 to 500 megatons/year. Robinson and Robbins (1971) estimated this input at 200 megatons/year (Mt/yr) from calculations based on steady-state concentration of atmospheric dust. This is in agreement with the 250 Mt/yr estimated by Peterson and Junge (1971) from extrapolation of data on the amounts of natural and agricultural wind-blown dust over the United States.

The uncertainty of particulate emissions from forest fires and slash burning are greater than those of other emissions and reflect the need for more definitive studies. An evaluation of man's contribution relative to those of natural processes is particularly desirable since it may be substantial and is amenable to control. The lower figure of 3 Mt/yr is the estimate of Robinson and Robbins and is based upon global extrapolation of assumed particulate combustion products in forest fires in the United States. Flohn (1971, unpublished) calculated 80 Mt/yr as the input from man-ignited bush fires, mainly in African savannahs. Robinson and Robbins also report an estimate by Hidy and Brock (1970) of 150 Mt/yr using different assumptions for particle production and global extrapolation factors of United States forest fires.

The total amount of salt injected into the atmosphere is estimated by Eriksson (1959) to exceed 1000 Mt/yr. Most of these particles are rapidly returned to the oceans, and only 100 megatons were expected to fall out over land. If it is assumed that the salt particles transported over long distances are represented by the land fallout, then the total fallout over the surface of the earth of such particles would be about 300 Mt/yr since the oceans represent about two-thirds of the earth's surface area.

On the basis of the rates of accumulation of montmorillonite in deep-sea sediments Goldberg calculated the volcanic debris emissions to the atmosphere to be of the order 150 Mt/yr. The lower value of 25 was made by Peterson and Junge (1971) by the extension of Mitchell's (1970) data to represent a long-term average.

The higher value of 200 Mt/yr for sulfate production from H_2S was taken from Robinson and Robbins. Peterson and Junge's estimate corrected for the conversion of gaseous emissions to sulfate alone results in 170 Mt/yr. Goldberg's lower estimate of 130 is based on the assumption that the conversion rate of biologically produced H_2S to sulfate plus the injection rate from volcanoes is the same as his computed value of the production rate of sulfate from fossil fuel combustion.

The upper and lower emission estimates for ammonium particles are from Robinson and Robbins and Peterson and Junge, respectively, again based on judgments of gaseous conversion rates. Robinson and Robbins estimate the 430 Mt/yr and Peterson and Junge the lower values of 60 for nitrate production rate.

Both the upper and lower limits of the range of hydrocarbon production from vegetative sources were derived by Junge and Peterson as well as by Robinson and Robbins from the data of Went (1960).

About 18 megatons of particles/yr are injected into the atmosphere from stationary sources in the United States (Shannon et al., 1970). The fine-particle component (radii less than 2 μ) is estimated to be about 22 percent of this figure, 4 Mt/yr. Goldberg assumed the world inputs were 3 times this value to give a global value for the fine-size fraction of about 12 Mt/yr. Robinson and Robbins give a total particulate emission of 90 Mt/yr for the entire earth.

Goldberg obtained the rate of formation of sulfate in the atmosphere following fossil fuel combustion by comparing present sulfate content of glaciers with those of time periods apparently unaffected by man. His value is 130 Mt/yr. Peterson and Junge give a value of 200 Mt/yr by modification of the data of Robinson and Robbins.

The nitrate pollution production rates were estimated from the composition of fossil fuel combustion products by Peterson and Junge as well as by Robinson and Robbins. They arrived at figures of 35 and 30 Mt/yr, respectively, not significantly different.

Peterson and Junge took their pollutant hydrocarbon emission number from NAPCA (1970), which gives a value for total hydrocarbon releases. They assumed that of the 29 megatons so released 75 percent is not methane and could be converted to aerosols. The SCEP Report gives 90 megatons of hydrocarbons released annually, perhaps an upper limit to particulate production.

In summary, it is apparent that the current state of knowledge on the emission and formation rate of particulate matter into the atmosphere is most unsatisfactory. The extreme limits of the indicated ratio of man-made particulate contribution to total particulate load is between about 5 percent and 45 percent, and a precise figure of acceptable reliability cannot be arrived at this time. The most important contribution by man is to the sulfate particles; this is now about 50 percent. If we assume from energy estimates that as of 2000 A.D. the production of man-made particles will be doubled, the corresponding figures would be 20 percent to 50 percent for the total man-made fraction and 70 percent for sulfate.

8.3.2
Formation of Particles

It is now believed that perhaps the majority of particles in suspension in the atmosphere are secondary products formed from material which entered the atmosphere as a gas. Several atmospheric gases, including SO_2, H_2S, and NH_3, are of importance in this process. Their residence time as gases in the troposphere is short, varying between a few days and about two weeks (in highly polluted areas perhaps only hours), because of their fast reactions with other components. As a result steady-state conditions are quickly established even if emission rates increase. Photochemical gas reactions that lead to the formation of particles are known (see also the section on NO_2 and hydrocarbons). The reactions between NH_3 and SO_2 are discussed here since these gases are important sources of secondary particulate matter on a global scale.

SO_2 has been measured extensively in polluted areas for many years. The emission of pollutant sulfur has been estimated in Section 8.3.1. Recent measurements of SO_2 in unpolluted air show that this gas is distributed on a global scale. Georgii (1970) and Büchen and Georgii (1971) have conducted an extended analysis of the horizontal and vertical distribution of SO_2. An evaluation of monthly SO_2 data by deBary and Junge (1963) showed a decrease of SO_2 concentration from the European continent toward the ocean. According to the present state of experimental technique, the background concentration of SO_2 seems to be 1 to 2 $\mu g/m^3$. This is also in agreement with results gained by Abel et al. (1969) at the observatory of Izana at 2370 meters altitude on the island of Tenerife. The SO_2 concentration in the pure maritime air over the Atlantic measured during the Atlantic expeditions of the research vessel *Meteor* shows a maximum at about 50° N with 3 $\mu g/m^3$ SO_2 and a decrease toward the north and the south. South of 10° N, SO_2 could no longer be detected over the Atlantic. It is to be assumed that SO_2 mainly originating in polluted areas is transported over long distances decreasing in concentration by slow photochemical and catalytic oxidation and by absorption at the ocean surface. The meridional profile presented in Figure 8.1 is in agreement with a residence time for SO_2 of 3 days and with the observation that there are no maritime SO_2 sources. With respect to the average vertical distribution of SO_2, aircraft measurement over central Europe showed

1. The vertical concentration of SO_2 decreases rapidly with altitude reaching half the ground concentration at about 1000 meters above the ground.
2. Seasonal variations of SO_2 can be observed only up to 2 kilometers.
3. Above 4 kilometers the SO_2 concentration does not decrease with increasing altitude. The measured continental SO_2 values

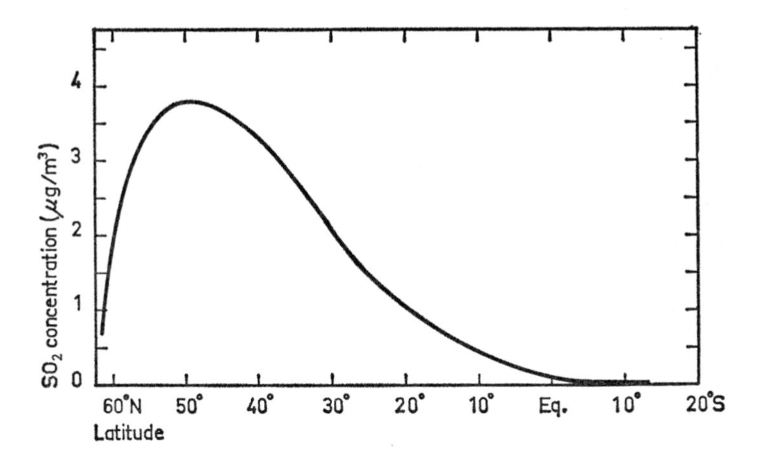

Figure 8.1 Meridional profile of SO_2 over the Atlantic Ocean.
Source: Data from the Faroer Expedition of 1968 and the Atlantic Expedition of 1969 of the German research vessel *Meteor*.

in unpolluted air over Europe are around 3 $\mu g/m^3$ and over the western part of the United States around 1 $\mu g/m^3$.

Direct measurements of H_2S are still extremely scarce and uncertain due to experimental difficulties. The predominant contribution to atmospheric H_2S comes from biological sources (decomposition of organic material) and from volcanoes. A series of H_2S measurements in Bedford, Massachusetts, by Junge (1960) showed a relatively constant concentration. Improvement of the analytical technique for H_2S deserves high priority. NH_3, also mainly produced by biological action and only to a small degree by combustion processes, is of importance in this context as partner in the formation of particles of ammonium sulfate, $(NH_4)_2SO_4$.

In relatively unpolluted air over Europe, NH_3 values of 5 to 7 $\mu g/m^3$ are found (Georgii, 1967). In the maritime air of Florida and Hawaii the concentration is 3 to 5 $\mu g/m^3$. Preliminary measurements of NH_3 by aircraft ascents in the free atmosphere indicate relatively small concentrations, around 1 $\mu g/m^3$. The vertical and horizontal distribution of NH_3 in the middle and higher layers of the troposphere deserves more attention in order to allow quantitative conclusions on the significance of these gases as sources of atmospheric aerosol particles.

8.3.3
Transformation of Gaseous Components to Particles

Sulfur dioxide is one of the principal air pollutants, and its oxidation to sulfuric acid is an important atmospheric process. The formation of ammonium sulfate particles is certainly significant to consideration of possible climatic effects since their presence has been established on a global scale.

Three mechanisms have been proposed for the formation of particles from gases in the atmosphere:

1. Photochemical oxidation, heterogeneous gas reactions.

This process is probably relevant only in arid areas and in higher layers of the troposphere. Cox and Penkett (1971) give a conversion rate of 0.03 percent SO_2 oxidized per hour in pure air by photochemical action. Bricard et al. (1968) have shown that if the SO_2 concentration in air is progressively increased, a saturation point is reached beyond which the production of particles is no longer a function of the SO_2 concentration. Irradiation of a mixture of $O_2 + N_2$ in the presence of traces of SO_2 did not give rise to particle production, but the addition of NO_2 and N_2 did. They concluded that the direct photooxidation of SO_2 alone does not form an appreciable number of condensation nuclei; the presence of NO_2 appears to be necessary for the nucleation (Bricard et al., 1971).

Photochemical reactions also occur with terpenes emitted from vegetable matter under the influence of sunlight. Over extensive forest regions these give rise to the phenomenon known as "blue haze." Went (1960, 1966) has studied this phenomenon extensively (see Section 8.6.1).

2. Catalytic oxidation in the presence of heavy metals.

According to Junge and Ryan (1958) this process depends largely on the presence of suitable catalysts—heavy metal ions—and is probably only effective in polluted air. They found that the transformation stops when a certain pH-value is reached. This mechanism most likely occurs both in dry particulate haze and in cloud droplets.

3. Ammonia-sulfur dioxide reaction in the presence of liquid water (clouddrop reaction).

The main features of the vertical and spatial distribution of sulfate particles in the troposphere and the concentrations can be understood when the SO_2-NH_3-liquid phase reaction is considered. The production rate is dependent on the NH_3 supply; if the pH-value is kept high enough—for example, by adding NH_3—the reaction can continue. According to our present knowledge, which may not be conclusive at this stage, this mechanism of ammonium sulfate formation is effective only in the presence of liquid water, that means regions where clouds and fog exist. Model calculations show an oxidation rate of 12 percent SO_2 per hour in cloud droplets. Recent aircraft measurements of individual particles containing $(NH_4)_2SO_4$ frequently show peak concentrations just below the cloud base. As a final product of this reaction, ammonium sulfate particles remain suspended in the atmosphere after evaporation of cloud or haze droplets.

When ammonia is added to filtered air and then enriched

with SO_2, there is an increase in the number of condensation nuclei from about $10^5/cm^3$ to $10^6/cm^3$ and a sharp increase in radius from about 3×10^{-9} to 10^{-6} cm (Bricard et al., 1971).

In summary, it would appear from experiments that the transformation of SO_2 into particles requires one or more other gaseous impurities in order to account for the observed effects. Since the average tropospheric residence of the $(NH_4)_2SO_4$ particles that are mainly found in the size range 0.2 to 0.8 cm radius exceeds that of the gaseous components SO_2, H_2S, and NH_3, it is concluded that this mechanism may contribute significantly to the global particle content and therefore have consequences for the atmospheric radiation budget.

8.3.4
Recommendations

1. We recommend the compilation of figures on global particle production rates that are accurate to a factor of 2 or better. For a number of constituents this will require the development of experimental and statistical methods to determine production or emission rates from, for example, slash burning, volcanoes, and entrainment of dust from overgrazed, arid, and desert areas. All studies should give information on mass fluxes as well as size distribution in the range between 0.01 and 10 μ radius.
2. We recommend that the transformations of atmospheric trace gases which lead to particle formation be studied. This study should include the collection of better data on the distribution of these gases in the atmosphere and on their life cycles and residence times.
3. We recommend periodic measurement (for example, at intervals of 2 years) of the major sources of particles which are man-made and over which he can exert control.

8.4
Removal of Particles and Trace Gases

8.4.1
Discussion

Natural and man-made substances are removed from the atmosphere by several processes. The following tropospheric sinks for particles and gas traces must be considered:

1. *Particles*

Dry deposition by sedimentation and diffusion.

Dry removal by impaction on plants.

Incorporation into precipitation elements by condensation, rainout, and washout.

2. *Gases*

Adsorption or reaction at the earth's surface.

Conversion into aerosols by chemical reaction within the atmosphere.

Incorporation into precipitation elements.

The efficiency of *impaction* of aerosol particles and molecules of gases depends on the rapidity with which pollution can be dispersed through the surface layers that constitute a resistance setting up a vertical gradient of pollution (Munn and Bolin, 1971). Some surfaces are almost perfect sinks for gases and particles; the substances are irreversibly absorbed. The other limiting case, perfect reflection, is assumed in diffusion models but probably does not occur in nature. The efficiency of impaction at the ground is still poorly understood; in particular the scavenging effect of forests is not yet predictable. Dew formed on the ground seems to be active in enhancing the deposition of particles and gases at the ground. The basic physical mechanism of these processes is not well known.

With respect to SO_2, one of the main gaseous pollutants, measurements of the deposition on different soils have shown that the absorption rate increases rapidly with increasing humidity and pH-value of the soil. Investigation of the depth of penetration has revealed that normally only a very thin surface layer contained SO_2 but for a very wet soil, transport into deeper layers of the ground was found. Taking the efficiency of absorption of charcoal for SO_2 as 100 percent, we find that the absorption efficiency of different soils varies between 20 percent and 90 percent, the highest values being for grass-covered ground. The deposition velocity, which is defined as the deposition rate divided by the air concentration, is about 1 to 2 cm/sec for SO_2.

Particles settle in the atmospheric gravity field. When transport by advection and convection is neglected, the sedimentation velocity determines the residence time of larger particles in the air. This process of *sedimentation* affects only particles larger than 1 μ radius.

The approximate sedimentation velocity in still air at 0°C and 760 mm pressure for particles having a density of 1 g/m³ is given in Table 8.2. From this table it can be seen that particles larger than 10 μ radius are quickly removed from the atmosphere by dry deposition. On the other hand, for particles smaller than 0.1 μ radius the settling velocity becomes negligible. The contribution of dry fallout to the total deposition rate in temperate latitudes amounts to about 10 to 20 percent. It can be taken for granted that in areas with little rainfall this percentage will increase. The onset of water vapor condensation on particles modifies the size distribution; particles grow and the rate of sedimentation increases.

Precipitation scavenging is one of the major processes of removal of gaseous and particulate pollutants from the atmosphere to the ground. It has three major aspects: delivery or transport of material to the scavenging site; in-cloud scavenging by the cloud elements, usually called rainout and snowout; and below-cloud scavenging by precipitation, usually called washout.

Rainout and washout are determined mainly by the following five parameters:

Size and concentration of the atmospheric particles.

Size and concentration of cloud and rain droplets acting as collectors.

Supply of liquid water by continuing condensation.

pH-value and chemical composition of cloud or rainwater.

Degree of solubility of gases and particles in water droplets.

Investigations by Beilke and Georgii (1968) and Beilke (1970)

Table 8.2
Dependence of Sedimentation Velocity on Particle Radius

Radius (μ)	Sedimentation Velocity (cm/sec)
0.1	8×10^{-5}
1.0	4×10^{-3}
10.0	0.3
100.0	25

Source: Junge, 1963.

Table 8.3
Contribution by Gaseous and Particulate Sulfur Components to SO_2 Concentration in Rainwater (percent)

	Sulfur Dioxide	Sulfate Aerosols
Rainout	5	20
Washout	70	5

Source: Beilke and Georgii, 1968.

show that in the case of SO_2: the rate of scavenging increases as the rainfall intensity increases (see Table 8.3); for constant rainfall intensity, the rate decreases as the droplet size increases; and the rate increases with increasing pH-value of the rainwater.

Computation of the relative contributions of washout and rainout to the SO_2 concentration in rainwater measured at the ground indicates that washout is the principal mechanism of removal in all cases when the SO_2 concentration is high near the ground and decreases with altitude. In a well-mixed atmosphere where concentration is independent of altitude, rainout undoubtedly predominates.

While the washout of trace gases is affected by Brownian diffusion, the incorporation of particles into cloud and rain elements is a result of both diffusion (in the case of Aitken nuclei) and impaction (in the case of large particles).

Theoretical calculations of the collection efficiency of particles by droplets have been carried out by Greenfield (1957) and others and were recently compiled by Engelmann and Slinn (1970) in the proceedings of a symposium on precipitation scavenging. Calculations on the individual contribution of aerosol particles containing sulfate and SO_2 show that in a polluted atmosphere the gaseous sulfur compounds contribute mainly to the sulfate concentration in rainwater as is shown in Table 8.3.

Several observations have shown that the removal of particles by clouds depends in some way on the type of precipitation. Itagaki and Koenuma (1962), Reiter (1961), and Georgii (1965) showed that snowflakes and falling ice crystals are more effective scavengers than rain droplets. Georgii attributed this effect to a high scavenging ability of the floating snowflakes with their large surfaces. Vittori and Prodi (1970) found that the capability of ice crystals (growing in a mixed cloud) to collect foreign particles is greater than that of condensing water droplets. Further laboratory investigations are needed to understand the role played by the diffusion processes (including electrical charge effects) that have been proposed by several authors as active mechanisms in transforming particles into droplets and ice crystals.

During and after individual rainfall and showers, a decrease of both the SO_2 concentration and the particle concentration near the ground is frequently observed. However, only cases with a rain intensity above 1 mm/hr lead to a noticeable decrease of the number of large and Aitken particles in the ground layer of the atmosphere. In the literature there are reports of cases when daily measurements of the turbidity showed a steady increase in spite of several showers which had occurred in the meantime (Flowers, McCormick, and Kurfis, 1969). It must be concluded that the production rate of trace substances in a highly industrialized region exceeds the trace substance removal by rainout and washout.

Reports on the heavy metal and lead concentration in precipitation by Lazrus (1969, unpublished data) and reports on the increasing acidity in Scandinavia and in other parts of Europe by Reiquam (1970) focus attention on precipitation chemistry and on the possibility to predict the cleaning of air by pollutant-collecting mechanisms active in clouds and in precipitation below the cloud base.

The removal processes largely determine the atmospheric *residence time* of particles and of gases reacting with precipitation elements. The residence time is defined by dividing the total mass in a reservoir (troposphere) by the total emission rate or removal rate (these are equal under steady-state conditions). Usually the residence time refers to the mass and not to individual particles that do not retain their identity in the process of coagulation.

In the troposphere the variation of size distribution and vertical concentration distribution will result in a certain variation of residence time, and some variation with latitude and with structure and frequency of clouds would also be expected. The tropospheric residence times of natural particles have in the main been calculated from data on atmospheric radioactivity. The residence times given by Junge (1963) fluctuate between 3 days and 22 days for temperate latitudes, a good estimate for ground air being 1 week.

The conclusion can be drawn that in temperate latitudes the wet deposition by fog and precipitation is responsible for about 70 to 80 percent of the mass removal from the tropospheric reservoir of particles and gases that react with water.

8.4.2
Recommendation

We recommend more comprehensive studies of the relative importance of the principal removal mechanisms of particles and gases by precipitation, of particles by impaction, sedimentation, and diffusion at the ground, and of gases by absorption at the ground. This has a particularly important bearing on computation of residence times in the atmosphere.

8.5
Particle Size Distribution

8.5.1
Three General Types

One of the basic parameters that enter all considerations of the effect of particles on radiation and cloud formation is their size distribution. On a global scale we can roughly distinguish three general types of particle distributions within the troposphere: "background," oceanic, and continental. Idealized average values that demonstrate the essential features of the three distributions are shown in Figure 8.2.

The most important particle distribution for global considerations is the one that we shall label "background." This particle population is representative for the middle and upper troposphere, that is, for about 80 percent of the troposphere. The background particle population is typical of what is usually considered very clean air, not affected by local or regional particle sources, either natural or man-made. It can be defined best by a lower limit in total number concentration as measured by a condensation nucleus counter. It seems reasonable to use 700 particles/cm^3 as this limit. The distribution in Figure 8.2 corresponds to about 200/cm^3.

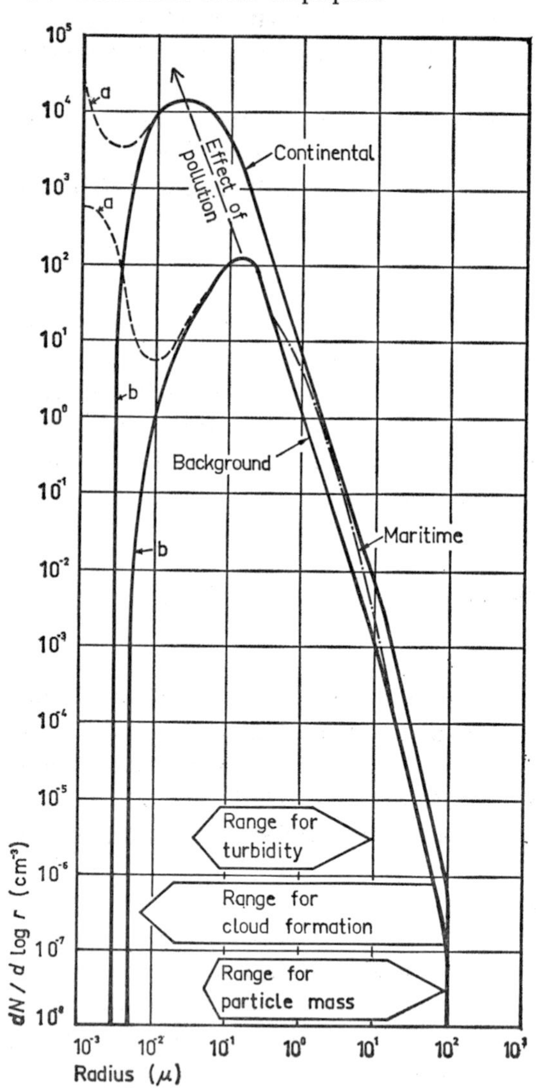

Figure 8.2 Typical comprehensive size distributions for the principle tropospheric regimes and the size ranges important for turbidity, cloud formation, and mass concentration of particles. Curves a and b refer to possible variation of the size distribution with and without continuous production of very small particles. The arrow indicates the effect of pollution on the location of the maximum of the size distribution.

Source: Data from Quenzel, 1970; Junge, 1955; Ikebe and Kawano, 1970; Junge and Jaenicke, 1971; as explained in Section 8.5.1.

Only very few data on this size distribution are available, with practically none below 0.1 μ radius. The combined evidence from optical studies (Quenzel, 1970), air electrical studies (Junge, 1955; Ikebe and Kawano, 1970), and the only direct measurements over the central Atlantic (Junge and Jaenicke, 1971) suggest that a maximum of the number concentration is about 0.2 μ radius (when plotted as in Figure 8.2 using the quantity $dN/d \log r =$ number of particles per cm^3 of air and per unit of $\log r$, for example, for the interval from 0.01 to 0.1 μ radius). With increasing radius the concentration decreases until a particle size of about $10^2 \mu$ radius. Almost nothing is known for sizes below 0.1 μ radius. Because small particles become very rapidly attached to larger particles by thermal (Brownian) coagulation, the concentration is expected to drop sharply with decreasing radius (as indicated by the solid line b of Figure 8.2, unless there is continuous production of very small particles. More recent laboratory studies have confirmed that sunlight, ionizing radiation, and the presence of trace impurities may result in such production (Bricard et al., 1968), and tentative data from the Atlantic Ocean (Junge and Jaenicke, 1971) suggest a distribution indicated by curve a, Figure 8.2, which indicates such production. The two curves, a and b, demonstrate schematically the possibilities with and without continuous production of very small particles in clean air.

The lowest layers over the oceans (about 2 km high) contain the marine particle population. This *oceanic size distribution* seems to differ from the background distribution only in the range between about 0.5 and 20 μ radius in which the sea spray particles produced by the breaking waves (Woodcock, 1953) are superimposed on the background distribution as indicated in Figure 8.2. The total concentration of these sea spray particles is rather small, less than $10/cm^3$, and their composition is similar to that of sea salt. Since the size range of particles that contribute most to the aerosol mass coincides largely with the sea salt range, marine aerosols as a whole have a composition dominated by sea salt.

In the lower troposphere over land, particularly in areas affected by pollution, we find the quite different *continental size distributions*. The total number concentration increases to about

$10^4/cm^3$ in rural areas, to about $3 \times 10^4/cm^3$ in towns, and to over $10^5/cm^3$ in cities (see, for example, Junge, 1963). The distribution in Figure 8.2 corresponds to about $10^4/cm^3$. Optical data (Bullrich, 1964; Yamamoto and Tanaka, 1969) indicate a uniform slope down to sizes of about 0.1 μ radius. Other direct measurements show that normally the maximum is not reached above 0.03 μ radius. Below a radius of 0.01 μ the size distribution is uncertain and the two possibilities are indicated by curves a and b, the latter, for example, indicated by recent data from Misaki, Ohtagako, and Kamazawa (1971).

8.5.2
Causes of the Distributions

There is evidence that particles larger than about 20 μ do not remain airborne for long because of sedimentation. The reason for the omnipresence of particles larger than this size, which is established (see, for example, Junge and Jaenicke, 1971) is not clear. The same is true for the shape of the whole size distribution. It is most likely that combined action of several processes is responsible for the fact that the important range above about 0.05 μ has a rather constant shape. One of the important factors appears to be the statistics of the size distributions of each of the large number of contributing sources, since particles normally travel once around the globe before they are removed from the atmosphere. In addition, there are several processes that constantly modify the size distribution, notably coagulation, which favors the attachment of small particles to larger ones. All these processes result in growth with time of individual particles that soon lose their identity and assume a very complex chemical composition. Table 8.4 demonstrates the effect of coagulation on the total particle concentration for a continental distribution similar to that in

Table 8.4
Change with Time of Total Particle Concentration and Mean Radius Due to Coagulation

Time (hours)	0	1	3	10	100
Concentration (cm^{-3})	27,000	20,000	16,000	9800	3500
Mean radius (10^{-2} μ)	3.2	4.0	4.7	6.0	10

Source: Junge, 1963.

Figure 8.2. Such processes as scavenging by rain, sedimentation, and impaction on the earth's surface are size-dependent and not only remove particles from the atmosphere but also continuously modify the size distribution.

Most atmospheric particles contain a substantial fraction of soluble material (about 50 percent on the average) that causes growth of the particles with increase of relative humidity (Kasten, 1968). This increase in size is substantial (more than a factor of 2) when the relative humidity approaches 100 percent. At the same time the particles assume more and more nearly spherical shape and their refractive index approaches that of water (see Section 8.7.1). Particle growth, therefore, affects turbidity and visibility considerably.

The wide ranges of size (about 5 orders of magnitude) and concentration (about 10) have to be considered with respect to their impact on radiation and formation of clouds and precipitation. In Figure 8.2 the relevant ranges are indicated. Extinction (turbidity) resulting in a decrease of direct solar radiation and visibility is caused by particles between about 0.05 and 10 μ radius. This range coincides closely with the range that controls the mass concentration of aerosols (that obtained by filtering the air). For this reason one observes a high correlation between visibility V (in km) and mass concentration M (in 10^{-6}g/m^3) expressed by $V \cdot M \approx 1200$, which is approximately valid for relative humidity below 75 percent (Charlson, Ahlquist, and Horvath, 1968). This correlation implies that the *shape* of the size distribution remains essentially unaltered above 0.1 μ radius.

The number concentration and average size of cloud droplets are strongly influenced by the number concentration of particles larger than about 0.02 μ (Squires and Twomey, 1960). The concentration and average size of cloud droplets in turn are important for the optical characteristics (that is, albedo) of clouds and fogs and hence for the atmospheric radiation budget (see Section 8.7.6).

The variation of particle size distribution with increasing pollution is important for any consideration of the effect of particles on radiation and cloud formation. Comparison of the background and continental distribution in Figure 8.2 demonstrates increase in total particle concentration, decrease of average particle size, and increase in total mass.

From measurements of electrical conductivity over the Atlantic Ocean since 1910 and other information (see Section 8.6.3), it can be estimated that the total number concentration in much of the Northern Hemisphere has *increased* by a factor of about 2, that the average size has *decreased* by a factor of about 1.5, and that the mass concentration has perhaps *increased* by a factor of about 1.5. All these figures must be considered very preliminary. In the Southern Hemisphere there are not yet any indications of global pollution by particles.

8.5.3
Conclusions

1. The general features of the particle size distribution are still very poorly understood, particularly in clean air and for particles smaller than 0.1 μ radius.
2. Different sections of the size distribution are responsible for the direct influence of particles on the radiation field and for the influence on the microstructure of clouds.
3. Monitoring of particles on a global basis for possible effects on climate should cover at least those size ranges responsible for these phenomena.

8.5.4
Recommendation

We recommend that suitable methods be developed to measure the particle size distribution below 0.1 μ radius and to study its modification in clean and polluted atmospheres.

8.6
Spatial Distribution and Trends of Particle Concentrations

8.6.1
Horizontal Distribution

Particles are rather short-lived constituents of the atmosphere with lifetimes ranging from a few days in layers near the surface to a few weeks in the upper troposphere (see Section 8.4). As a result spatial and time variations are rather pronounced and can amount to more than an order of magnitude. It is useful to distinguish areas with large particle concentrations due to local sources, regions with generally larger concentrations due to higher particle production over areas of the order of 1000 km, and areas on a hemispheric scale. There is very good evidence that the Northern and Southern Hemispheres are, as far as particles are concerned, completely independent. The average exchange time between the two hemispheres is about 1 year. Since the average lifetime of particles is shorter than about 3 weeks, it can easily be calculated that much less than 1 part in 1000 of the particulate material (and, of course, of all trace constituents of similar lifetimes) of one hemisphere can enter the other. It has to be considered, however, that in this context the hemispheres are separated by the meteorological equator (normally called ITC), which does not coincide with the geographical equator and varies in position with the season, most conspicuously over the northern Indian Ocean during the Indian monsoon.

Horizontal distributions are determined by the source and sink areas. Figure 8.3 gives a good example of the regional distribution of the total tropospheric particle load of the United States (Flowers, McCormick, and Kurfis, 1969) as revealed by tur-

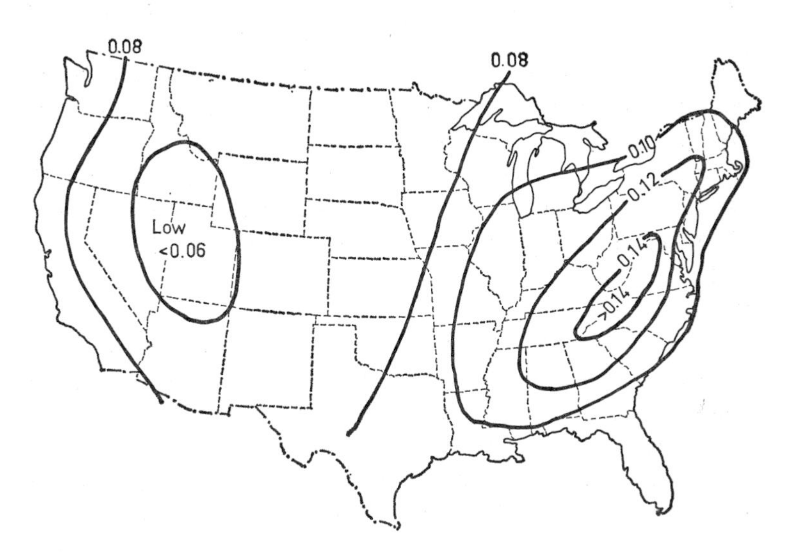

Figure 8.3 Average distribution of turbidity over the United States. Source: Flowers, McCormick, and Kurfis, 1969.

bidity data. The broad maximum over the Appalachians is assumed to be due to natural sources (transformation of terpenes to particles), and high values in the densely populated areas of the Northeast are attributed to anthropogenic sources. Similar patterns must be expected for other constituents. This demonstrates how difficult it is to delineate areas affected and not affected by natural or man-made pollution and how dangerous it is, unless the general distribution is known, to draw conclusions on global effects from a few isolated measuring stations even if these stations are considered to be representative of clean air conditions. Figure 8.3 is, of course, based on stations that are not influenced by local sources. In Figure 8.3 one must imagine that the areas of local pollution with diameters of 10 to 100 kilometers are superimposed on the regional distribution as relatively small spots with rather high values.

The large-scale distribution of particulate matter is mainly controlled by the large-scale wind systems. Because of the prevailing west winds in middle latitudes, particles produced over the North American continent are carried over wide distances across the Atlantic Ocean, as demonstrated by the amount of mineral dust and fly ash particles found on cruises between northeast United States and Ireland (Parkin et al., 1970). Periodic occurrences of Sahara dust carried with easterly trade winds all the way over the Atlantic Ocean are also well documented (see, for example, Robinson and Robbins, 1971). Haze occurrence over other oceans and widespread eolian dust found in deep-sea sediments attest the importance of particle transport over considerable distances through the atmosphere (Goldberg, 1971).

8.6.2
Vertical Distribution

If we exclude sea spray, all major direct sources for particles, both natural and man-made, are over land. For this reason there is generally a pronounced decrease in particle concentration with increasing altitude over land or over the adjacent ocean areas. The vertical profiles of particle concentration vary considerably with the meteorological conditions, but data indicate that at about 4 to 5 kilometers the influence from below becomes negligible

(Junge, 1963). Even over the center of the North American continent total particle concentrations above 5 kilometers were consistently found to be about 300/cm^3. This is the result of the large ratio of horizontal to vertical transport in the middle troposphere, particularly in middle latitudes. On the basis of the available data, we can therefore expect background size distributions (see Section 8.5.1) above about 5 kilometers.

Over the low-latitude oceans, sea spray particles seem to be restricted to the lowest 1 to 2 kilometers (Woodcock, 1953). There are almost no data on the vertical distribution of sea salt particles in middle latitudes, but it seems that the concentration drops very rapidly above about 2 kilometers even in the areas of intense vertical mixing.

8.6.3
Trends

Unfortunately there are only a few series of observations long and consistent enough and in places not much affected by local or regional pollution to give information on global changes of turbidity. From data in many places in the U.S.S.R. it is obvious that apart from decreases of direct solar radiation of several percent after major volcanic eruptions in 1903, 1912, and 1963 there is a systematic decrease over the last 25 years amounting to about 5 percent at a sun height of 30° in places remote from cities. A similar trend that is likely to be related to widespread pollution is indicated for clean air stations over southeast Japan and the western Pacific for the period 1948 to 1955, when measurements were terminated (Kitaoka, 1959) as well as at Mount Wilson Observatory during subsidence conditions since 1910 (Hodge, 1971). However, it is difficult to separate regional and global effects at the present time and to give reliable global average values of turbidity increase.

Figure 8.4 shows a set of data on the electrical conductivity of air which clearly shows that over the North Atlantic there was a continuous decrease through this century, whereas over the South Pacific there was no change (no data are available for the North Pacific). The observed decrease in electrical conductivity is caused by an increase by a factor of about 2 of the 0.01 to 0.1 μ radius

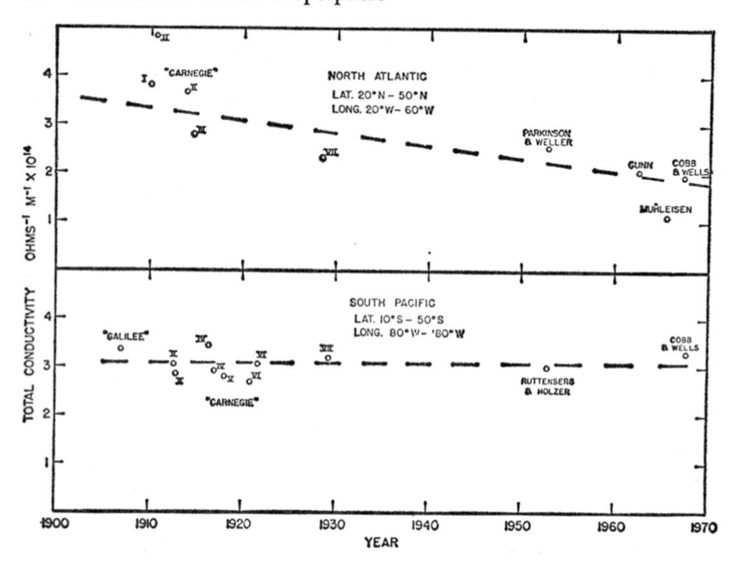

Figure 8.4 Data from electrical conductivity measurements over the North Atlantic and the South Pacific.
Source: Cobb and Wells, 1970.

particles (Cobb and Wells, 1970). This set of data is particularly important because it refers to an undisturbed area and may therefore be representative for background conditions.

From the existing data on turbidity it is difficult to derive a reasonably reliable figure for the whole Northern Hemisphere because of the large areas over the oceans where data are almost completely missing and because of the regional variability of the data over land. By combining the available evidence, an increase up to about 50 percent in the turbidity coefficient for the areas of low turbidity may be indicated for the Northern Hemisphere, but it should be kept in mind that this value may be less representative south of 30° N, where data are very scarce. A value of less than 50 percent would be consistent with the estimates on emission rates given in Table 8.1. No data from the Southern Hemisphere are available to our knowledge, but it seems unlikely that a noticeable increase has occurred.

In summary, our knowledge about the regional and global horizontal distribution of particles is very unsatisfactory. For the vertical distributions the situation is not much better, and data on regional or global increases over the last decades are too few or too unreliable to allow any firm conclusions.

8.6.4 Recommendations

We recommend the spatial distributions of particle concentration and trends with time be monitored on a global basis. A network of about 100 stations is required to give representative data for the whole atmosphere. For all optical measurements an accuracy of 5 percent is required to obtain meaningful data.

Because of the wide range of particle sizes, different methods have to be used simultaneously, in particular:

a. Trends of the total atmospheric load of particles in the size range above 0.1 μ radius should be monitored by determining the transmissivity of the atmosphere using a standard sun photometer that operates, for example, at wavelengths of 5000 and 3800 Å. At observatories more sophisticated and absolute instrumentation capable of spectral resolution in narrow bands (for example, 100 Å) in the ultraviolet and visible regions is recommended.

b. Horizontal extinction should be monitored at selected stations by transmissometers at selected wavelengths in water vapor windows. These measurements should be supported by measurements of the total mass concentration of particles by using filtering techniques. The filter samples allow chemical analysis that is required at intervals.

c. Trends in the concentration of particles in the size range below 0.1 μ radius should be monitored because this range is particularly sensitive to pollution and has considerable importance for cloud formation. The best choice for continuous monitoring is by a condensation nucleus counter with suitable diffusion or electrostatic filters for the proper size range. Stations with long records of electrical potential gradient, conductivity, and small ion concentration should continue such measurements.

d. For ocean areas continuation of periodic (for example, every 5 years) measurements of electrical conductivity and related data are recommended in order to keep up the existing long series of such observations.

e. Trends in the concentration of ice and cloud nuclei should be continuously monitored. Very careful selection and perhaps improvement of existing methods are necessary.

8.7
Particles and the Radiation Field
8.7.1
Refractive Index

The interaction of a given particle with the radiation field depends upon the shape of the particle and upon its refractive index m and its index of absorption k; m and k vary with wavelength of the radiation. For a spherical particle the dependence on shape is reduced to a dependence on the radius of the sphere. If the particle changes its composition or if its size changes (for example, by condensation of water), then m and k may change also. Because it is mathematically convenient, m and k are usually used in combination, the combination being called the complex refractive index, with a real part m and an imaginary part depending on k. Both parts determine the scattering of the particles; the index of absorption determines the absorption of the particle. The coefficient of absorption a_λ, which is used in Beer's law, is related to the index of absorption k by

$$a_\lambda = \frac{4\pi}{\lambda} k_\lambda$$

Sometimes the use of the coefficient of absorption is preferred, but in connection with scattering theory the index of absorption is very often used. Our figures are given in terms of the index of absorption. As we have only little information about the shape of atmospheric particles, it seems the best to consider the optical properties as those of a sphere with the same volume as the particle. The radius of this sphere is called the equivalent radius.

Soluble particles increase in size by adsorption and condensation of water vapor as the relative humidity increases. Thus with increasing humidity the index of absorption and refractive index of the particle approximate more and more to those of water. Under normal conditions atmospheric particles consist of about

50 percent soluble material. The variation of the radius with increasing humidity can be computed.

Since the refractive index of water is smaller than that of dry particles in the visible region, the index of refraction decreases with increasing relative humidity to the value 1.33. In the visible region, the index of absorption for water is small (10^{-7}) compared with that of the particles (~ 0.01). But in the infrared region the index of absorption of water is large (0.04 to 0.4) and most likely a little larger than that of the particles. Thus in the visible region the index of absorption goes to zero when the relative humidity reaches 100 percent, but in the infrared region it goes to about 0.04 to 0.4 when the relative humidity reaches 100 percent.

Only a few observations of the refractive index of particles are available, and we are able to give the refractive index and the index of absorption for one industrial location only. The refractive index m was found for that industrial area to be (Hänel, 1968; see also Deirmendjian, 1969) $m = 1.57$, valid for the wavelength 0.57 μ and for a mean density of 2.7 $\mu g/cm^3$ at a relative humidity of 40 percent. The variation of the refractive index with humidity for industrial particles may then be computed.

For the same location, a very tentative determination of the index of absorption for the airborne particles by optical means (Eiden, 1966) resulted in values of $k = 0.01$ to 0.1. Probably the higher values are valid for very heavily polluted areas. The latter fact was established in the laboratory by a determination by optical methods of the index of absorption of particle films obtained in the same area by filtration. It was measured for a relative humidity of 35 percent and the variation of the index of absorption was then calculated (Fischer, 1970). These results are given in Figure 8.5 for weakly absorbing material and for strongly absorbing material for three different wavelengths. The index of absorption increases with increasing wavelength, but the coefficient of absorption a_λ is nearly independent of the wavelength; that is, the absorption of these samples of continental particles is gray in the visible region.

In summary the index of absorption and the index of refraction are known for very few locations and even there only for the

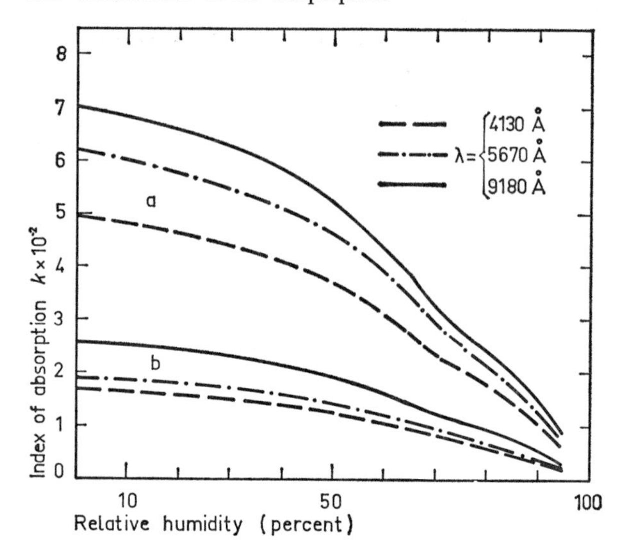

Figure 8.5 Variation of the index of absorption with relative humidity (a) for strong-absorbing particles and (b) for weak-absorbing particles (given for three wavelengths).
Source: Fischer, 1970.

shortwave spectral region. Knowledge of both is needed to calculate the extinction of radiation due to the particle layer and the heating rate of particle layers caused by the absorption of direct and scattered solar radiation and the absorption and emission of the particle layer in the infrared region.

8.7.2
Direct Effects on the Radiation Field

The climatological importance of radiative transfer and the scattering, absorption, and emission of radiation by particles have been discussed in Sections 5.2.1 and 5.2.2.

The scattering, absorption, and emission of an element of volume of the real atmosphere represents the integrated effect of the gas molecules and of all the particles, not all identical, contained in it. We shall consider separately how particles can change the solar and longwave infrared radiation in the atmosphere. In this section we shall not consider in detail the interaction of particles and clouds. This separation is rather artificial but very convenient (see Section 8.7.3).

PARTICLES AND SOLAR RADIATION

In the short-wavelength range, particle layers in the atmosphere may change both the global albedo and the absorption of radiation by the atmosphere. A full analysis of the effect of atmospheric particles on the transfer of radiation through the atmosphere is very complicated. In many contexts, simple idealized models are helpful. The simplest but still useful model envisages upward and downward streams of radiation that are considered absorbed and reflected. The fraction of the radiation that is reflected is often called the "backscatter." The ratio of the energy absorbed in the layer to that reflected determines, together with the albedo of the underlying surface, the heating or the cooling of the atmospheric layer (see Section 5.2.1).

The first applications of simple radiative transfer models to the problem of actual atmospheric particles proved inconclusive, and within the last year or two several more complex analyses have been made. Yamamoto and Tanaka (1971), Rasool and Schneider (1971), and Korb and Zdunkowski (1970) have incorporated the effects of multiple scattering and have arrived at very similar conclusions, which we illustrate here by reference to the work of the first-named authors. They assumed the following: the size distribution of Junge, the height distribution of Elterman (1964), the refractive index of particles of 1.5, and the absorption index of particles of 0 to 0.01. Using these assumptions, they computed the change of average global albedo with change of total number of particles for different surface albedos. Some of their results are shown in Figure 8.6 where global albedo is plotted against a turbidity coefficient which *with the above assumptions* depends only on total particle number.

Figure 8.6 indicates that for this specification of the particle load an increase in particle load increases the global albedo and so must cool the planet. Figure 8.6 also illustrates results for an atmosphere of average cloudiness above a surface of average albedo, obtained from a more complicated model. This is the most realistic case studied by Yamamoto and Tanaka, and the results are perhaps the most convincing evidence available now that an increase in the number of atmospheric particles would tend to cool the planet.

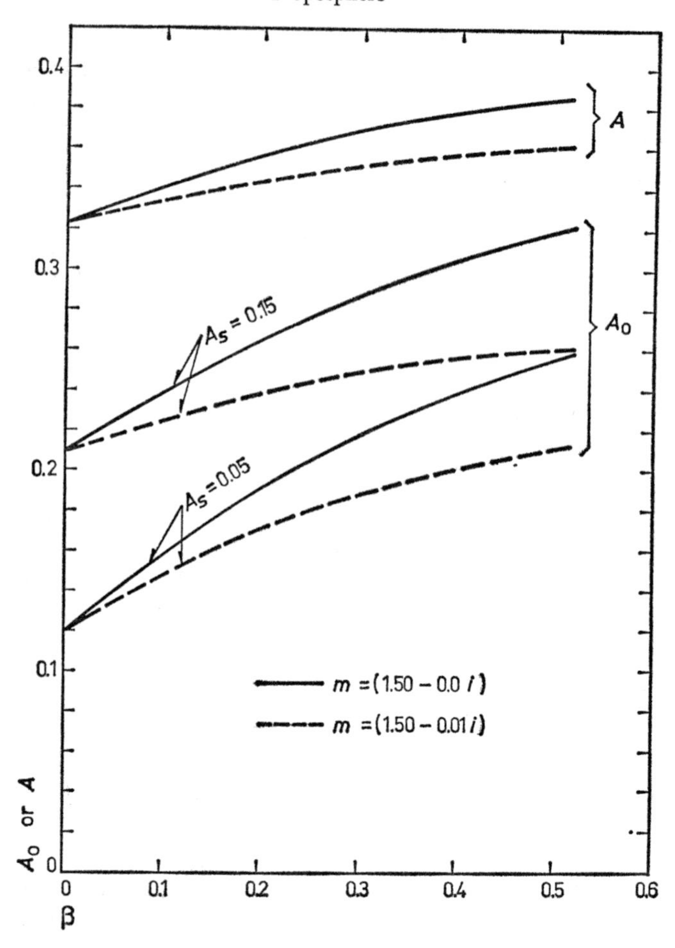

Figure 8.6 Global-average albedo in a clear atmosphere A_0 and in a cloudy atmosphere A as a function of β. The relation $A = nA_c + (1 - n)A_0$ is used, where A_c is the albedo of clouds, A_0 is the albedo of cloudless but turbid atmosphere, A_s is the surface albedo, m is the complex refractive index, and n is the fraction of the sky covered by cloud.
Source: Yamamoto and Tanaka, 1971.

It must be remembered, however, that these computations are made with an assumed refractive index, and we have seen (Section 8.7.2) how few actual measurements of the refractive and absorption indexes of atmospheric particles have been made. An alternative test of the assumptions can be made by measuring the upward and downward streams of solar radiation above and below a layer of particles (or an artificial radiation source can be used). Figure 8.7 (Robinson, private communication) illustrates some results of this method. Measurements of the absorption and "back-scattering" coefficients of atmospheric particle layers are plotted on a diagram (based on the model of Atwater, 1970) which show for two different values of surface albedo whether the layer would increase or decrease the global albedo. The diagram suggests that the particles characterizing heavy industrial pollution might decrease global albedo over many land surfaces. More measurements of the properties of the particles are clearly required.

Yamamoto and Tanaka also computed the decrease in the effective radiative temperature of the earth corresponding to their computed albedo change. This is shown in Figure 8.8 as ΔT_e, which follows directly from the planetary heat balance. They also used an empirical global-average model of Budyko (1969) to compute the corresponding change of surface temperature, and their results are plotted as ΔT_{ste} in Figure 8.8. These results for surface temperature are subject to all the caveats made elsewhere in this volume concerning the indications of both empirical parameterized models and global-average models (see Section 6.4). Nevertheless, they demonstrate how important it is to know the optical properties of the particles with which man loads the atmosphere.

PARTICLES AND INFRARED RADIATION

The effective radiative temperature of the planet depends only on the amount of solar radiation retained. The properties in the far infrared of any substance likely to be found in a clean or polluted atmosphere do not affect this temperature. These properties do, however, directly affect the temperature distribution in the atmosphere: particles can conceivably contribute to the "greenhouse effect," warming the surface at the expense of compensating cooling higher in the atmosphere. The actual effect observed on

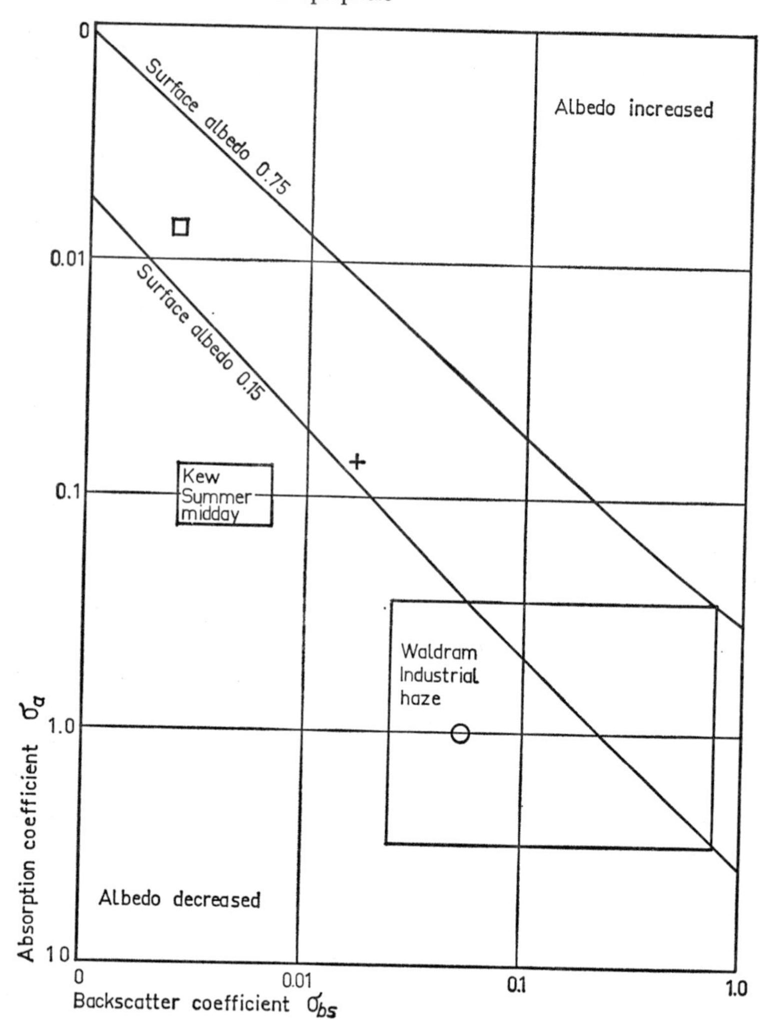

Figure 8.7 Absorption coefficient versus backscatter coefficient of a particle layer for two values of surface albedo. If the characteristic point lies to the right of a surface albedo curve, the total albedo is increased.
Source: Robinson, 1971.

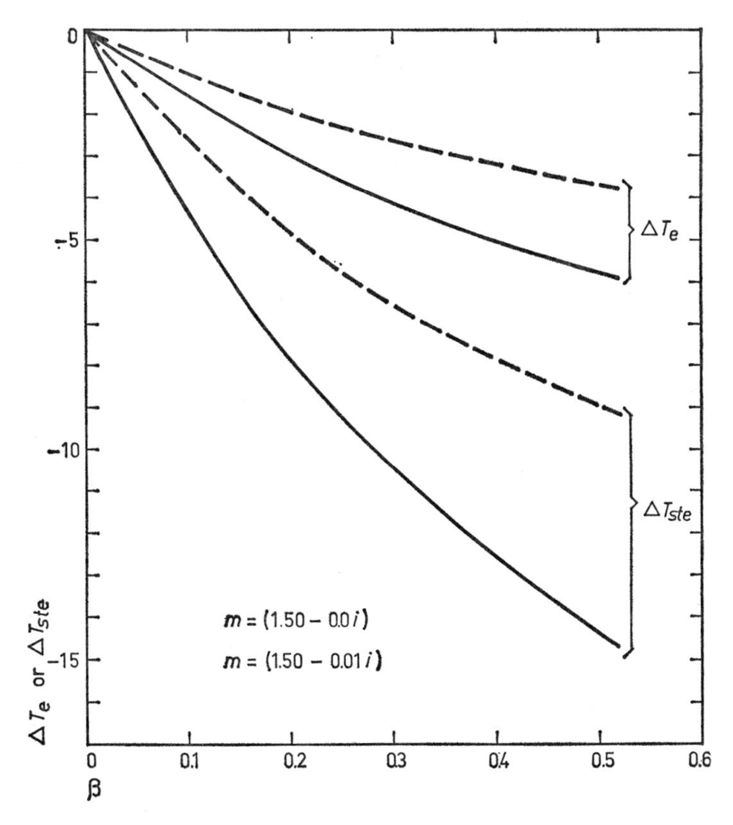

Figure 8.8 The decrease in surface and effective temperature as a function of atmospheric turbidity expressed by β.
Notes: The effective radiation temperature is defined as

$$T_e = \sqrt[4]{\frac{S(1 - A)}{4\sigma}}$$

where S is the solar flux, A is the planetary albedo, and σ is the Stefan-Boltzmann constant. The surface temperature is found from

$$\pi R^2 \cdot S(1 - A) = 4\pi R^2 I(T_{ste})$$

where R is the radius of the earth, T_{ste} is the surface temperature, and

$$I(T_{ste}) = a + bT_{ste} - (a_1 + b_1 T_{ste})$$

is the ground emission as given by Budyko (1969).
Source: Yamamoto and Tanaka, 1971.

any occasion depends in a complicated way on the actual temperature structure of the atmosphere. Several investigators have reported results of this type of computation (Atwater, 1971; Joseph, 1971; Kattawar and Plass, 1971; Rasool and Schneider, 1971; Korb and Zdunkowski, 1970); and the general outcome is an indication that the effect of atmospheric particles on the infrared radiation field is of less importance than that on solar radiation, and that the "greenhouse effect" of particles is unlikely to compensate, except in very heavy industrial pollution and at high latitudes, for the surface cooling resulting from the attenuation of solar radiation.

However, since the surface albedo and the size and number of particles in the atmosphere are highly variable over the earth, global-average models may not be of sufficient resolution to estimate the overall effect of particles on the global albedo. Modeling the effects of particles on the radiation balance should be performed with models of sufficient complexity to account for the distribution of particles, surface albedo, cloudiness, and incoming solar radiation over the globe (see Section 6.5.4).

8.7.3
Modification of Cloud Properties

THE FORMATION OF CLOUDS

All cloud particles are originally formed by the condensation of water substance on a particle. Clouds may be water droplets, ice particles, or a mixture of both. It is necessary to distinguish among them since a particle that is an efficient nucleus in one process is not necessarily efficient in another.

Water Clouds

Water clouds are formed above 100 percent relative humidity (R.H.) by condensation upon suitable particles. Since the act of condensation depletes the vapor content, the relative humidity does not rise much above 100 percent before it begins to fall again. For this reason many, usually most, of the particles in the air do not become cloud droplets.

The exact relative humidity attained in a given situation depends on the rate of cooling and on the numbers of nuclei becoming activated at various values of the humidity once 100 percent R.H. is exceeded. However, the number activated does not vary rapidly with relative humidity, and under such circumstances it can be demonstrated that the dependence upon cooling rate is weak; the dominant effect is that of the particles that are active condensation centers below 101 percent relative humidity. The value of 101 percent R.H. represents the upper limit under most actual conditions in the atmosphere. The particles that have been activated below 101 percent determine how many cloud droplets are formed, and these are referred to as cloud nuclei. They are measured by applying relative humidities of 101 percent or less to a sample volume of air and counting the droplets formed.

The relationship between cloud nucleus concentration and cloud-droplet concentration has been substantiated by direct comparison (Squires and Twomey, 1960, Twomey and Warner, 1967) and appears to be reasonably well established; that is, from a measurement of cloud nuclei numbers we can predict cloud droplet numbers with some confidence.

Mixed Clouds

Some particles have been found to have the properties of initiating the freezing of a cloud droplet that contains them; they are called ice nuclei. Ice particles begin to appear in a parcel of cloudy air when its temperature is sufficiently low. Extensive measurements of ice nuclei have been made (for example, Bigg, 1965; Bigg and Stevenson, 1970), but at the present time there exists a large and unexplained discrepancy between the number of ice nuclei and the usually much larger number of ice particles actually found in clouds (Köenig, 1963, 1968; Mossop, Ruskin, and Heffernan, 1968). Until the discrepancy has been resolved, the number of ice crystals in a cloud cannot be predicted with confidence.

Ice Clouds

Ice clouds may be formed by glaciation of a liquid water cloud—a process that can be so fast as to appear instantaneous at cirrus cloud temperatures—or by sublimation (growth of ice from the vapor phase without initial formation of liquid water). The first-mentioned process is believed to be dominant. However, the most obvious aspect of cirrus is our lack of even quite rudimentary information on the subject.

Once formed, ice particles will survive below 100 percent relative humidity—in fact will continue to grow—until the humidity falls below that value which represents ice saturation. Table

Table 8.5
Relative Humidity for Ice Saturation

Temperature (°C)	Relative Humidity (percent)
−10	91
−20	82
−30	75
−40	68

Note: The relative humidity figures were derived from Tables 108 and 109 of the *Smithsonian Meteorological Tables.*
Source: List, 1963.

8.5 gives the relative humidity for ice saturation at several temperatures. The consequence of the ice-water vapor-pressure difference can therefore imply that once cirrus forms, even in a thin layer, it may develop throughout the ice-saturated layer. This behavior, together with the well-recognized influence of cirrus, which is discussed elsewhere in this report (see Section 8.7.6), emphasizes the need for much greater effort in the study and observation of cirrus clouds.

NUCLEATION AND THE PROPERTIES OF CLOUDS

Cloud properties can be influenced by particles in several possible ways:

1. Additional cloud nuclei raise the drop concentration and therefore (other things being equal) reduce the average size of the drops.
2. Addition of ice nuclei increases the concentration of ice crystals in a cloud at temperatures below zero provided the cloud is not already completely glaciated.
3. Addition of sublimation nuclei can also give rise to cloud formation in air above ice saturation even if it is below water saturation.
4. Addition of very efficient large soluble particles before cloud formation occurs can in principle deplete the vapor content so that the humidity is held down and fewer cloud nuclei become activated than otherwise.

At the present time the third and fourth effects listed can be discounted, the third because of the rarity, even perhaps absence, of sublimation nuclei in the atmosphere, and the fourth because of the very special combination of circumstances needed to make possible an appreciable change in the number of cloud drops by that mechanism.

MAN-MADE CLOUD AND ICE NUCLEI

Examples of the modification of ice nucleus concentration by man-made emissions, especially from steel manufacturing plants, are numerous in the literature (Soulage, 1958; Telford, 1960). The issue is, however, confused through the aforementioned discrepancy between ice-particle concentration in clouds and measured concentrations of ice nuclei. The addition of freezing nuclei is, of course, the basis for seeding supercooled clouds in rainmaking experiments, but it is also a mechanism that may change drastically the microstructure of cloud, from perhaps very many small liquid drops to much fewer but larger ice particles.

The effect of the first-mentioned mechanism has been observed both by direct measurement of clouddrop concentration and by measurements of cloud-nucleus concentrations. Warner and Twomey (1967) have described observations of increased numbers of cloud nuclei and clouddrops caused by sugarcane burning in Queensland. Increased cloud nucleus counts have been found in air that has experienced urban-industrial pollution. A tenfold increase in nucleus concentration several tens of miles from the source is quite common, but on a global scale only a few percent of the cloud nuclei are estimated to have been injected into the atmosphere *in particle form* by man's activities (Squires, 1966; Twomey and Wojciechowski, 1969). However, the main source of cloud nuclei is probably particle formation in the atmosphere from trace gases, and the nuclei will therefore be influenced also by gaseous emissions (see Section 8.3.3). The extent of man's indirect influence cannot be gauged at our present state of knowledge.

Since late 1968 systematic measurements of the concentration of cloud nuclei in Australia have not disclosed any increasing trend. The period covered is still too short for anything other than gross trends (greater than 10 percent per year) to be perceptible. Measurements at different geographic locations show a pronounced influence of the continents upon the cloud nucleus con-

Table 8.6
Typical Concentration of Cloud Nuclei (measured at 101 percent R.H.)

	(cm^{-3})
Over continents	400
North Pacific	100
North and South Atlantic	100
Southern ocean	40

Source: Twomey and Wojciechowski, 1969.

centration (Table 8.6), but the main continental contribution does not appear to be dependent on population and does not seem to be directly due to man's activities.

With its low numbers of cloud nuclei southern ocean air would be most sensitive to a global buildup of cloud nucleus numbers. It is suggested that monitoring programs, whether by direct sampling or indirectly related observations (such as shortwave albedo or cloud brightness measurements from satellites), should not overlook this aspect.

The present global production of cloud nuclei has been estimated (Twomey and Wojciechowski, 1969) to be about 10^{26} nuclei per day or about 4×10^{28} nuclei per year. The average mass of the nuclei is of the order of 10^{-17} g, so that about 4×10^{5} tons is estimated for the mass of cloud nuclei produced per year. This is quite a small mass compared with, for example, the 80 million tons of sulfur that Junge estimates (see Section 8.3) to be put into the atmosphere annually by man's activities. It is clear that even with present emission rates there is far more material in the atmosphere than is needed for the production of the cloud nuclei.

8.7.4
Modification of Cloud Cover

It has been demonstrated theoretically that a cloud containing many droplets will, because of the small size of the droplets, not produce rain as readily as a cloud containing relatively few, and therefore larger, droplets. In this way an increase in the number of cloud nuclei could lead to a decreased efficiency of rain formation, especially in regions where rain is formed predominantly in nonsupercooled clouds and where the air is naturally low in cloud nucleus content. If the water vapor input remained constant an *increase* in cloud cover or cloud depth (or both) would result. The extent to which this inhibition of rain formation occurs in the real atmosphere is not known. Further research is needed to clarify the point.

Another possible effect of particle pollution works in the opposite sense: if giant soluble particles or effective ice nuclei are added inadvertently or otherwise by man's activities, then a possible stimulation of precipitation or cloud dissipation, accompanied by a *reduction* in cloud cover could result. However, giant soluble particles and effective ice nuclei are much more special kinds of particle than are cloud nuclei. The latter have been measured in greatly enhanced numbers in polluted areas (Squires, 1966), but there is no clear evidence that the former are produced in significant amounts by general pollution (SCEP Report, 1970).

To sum up, it may be said that mechanisms exist whereby cloud cover may be increased by pollution. Mechanisms that are probably less important tend to decrease cloud cover. Our present state of knowledge does not allow us to estimate the extent to which cloud cover has been altered or is being altered by the general worldwide increase in pollution. Precipitation may be influenced by cloud seeding that is artificial but not deliberate if it is stimulated by particles such as freezing nuclei released from steel manufacturing plants; silver iodide, or other artificial nucleating agents carried over from a distant cloud-seeding operation; giant particles or drops from industrial or agricultural sources; freezing nuclei emitted or formed upon exhaust emissions from various kinds of engines. These effects are believed to be at present no more than marginal. Furthermore, a feedback or similar mechanism would be needed to give rise to anything more than a local redistribution of precipitation amount. For both reasons the possibility of global climatic effects arising from these causes appears somewhat remote.

8.7.5 Pollution of Droplets

Pollution of droplets by soluble and insoluble materials may give rise to absorption of shortwave radiation. Since measurements have shown that absorption by clouds in the visible may be appreciable (Robinson, 1958) it is of obvious importance to clarify the question of where and how shortwave radiation can be absorbed within a cloud. The absorption coefficient of bulk pure water in the visible is too small to give any appreciable absorption by the water. Suggestions have been made that in thin films or small drops of the dimensions of clouddrops the absorption coefficient may be larger than for bulk water and that small ($\sim 0.5\ \mu$) absorbing particles within the clouddrops are responsible (Danielson, Moore, and Van de Hulst, 1968).

This important question can be resolved only by further measurements of the radiation fields, and careful laboratory measurements of the absorption coefficient of cloud-water samples are needed both in polluted and in cleaner regions and at several levels in the atmosphere.

It is a well-known fact that precipitation elements contain trace substances in wide range of concentration depending on the available liquid water content and the environmental conditions. The difference in the trace-substance concentration of a maritime fog in contrast to the London fog illustrates strikingly the problem of droplet pollution.

Trace substances—trace gases as well as particles—are incorporated by different mechanisms into cloud and fog droplets and are impacted by falling raindrops. Besides these mechanisms the liquid water of the drops also plays a role in providing a medium in which chemical reactions occur (see Section 8.3.3).

The electrolytic conductivity of liquid water, depending on the concentration of ions in the droplets, can be taken as an indicator of their state of pollution. Measurements by Georgii (1965) and by Mrose (1966) in raining and nonraining clouds in the Alps and on the Taunus ridge in West Germany as well as in fogs in Thuringia show clearly the amount of trace material accumulated in continental ground fog and near the base of clouds, and also, the decrease in concentration with altitude. High values near the cloud base are produced by transport of trace gases and aerosols from the polluted boundary layer of the atmosphere.

In rain clouds we can expect the trace-substance concentration to increase by 20 to 40 percent between cloud base and ground owing to pickup of material during the fall through the cloud-free zone and to partial evaporation. The results indicate that two-thirds of the trace content collected in rainwater at the

Table 8.7
Sulfate Concentration in Cloud Droplets

Altitude above Sea Level (m)	Sulfate Concentration (mg/liter)
550	0.75
850	0.7
1800	0.4
2300	0.4

Source: Georgii, 1965.

ground has been incorporated during cloud formation while one-third of the concentration is due to processes active below the cloud base. It is understandable that the ratio may vary considerably, depending on the degree of pollution of the surface layer of the atmosphere.

Measurements of the sulfate concentration in cloud water near Innsbruck, Austria, yielded the values shown in Table 8.7.

Oddie (1962) reported aircraft measurements near the base of clouds over southern England and found appreciable amounts of sulfate and other pollutants (4.1 mg/liter SO_4 on one occasion at 1950-meter altitude when the aircraft was downwind of the Birmingham area).

Results thus far obtained show that we can consider as heavily polluted only the layer clouds and ground fog in areas with a high production rate of pollutants. The higher the clouds reach into the atmosphere, the "cleaner" we can expect them to be. This is in full accord with our present knowledge on the vertical distribution of trace gases and aerosols. To support these findings further, it will be necessary to sample cloudwater at different altitudes within clouds of different type.

With respect to droplet pollution by organic substances, attention is drawn to the work on the pickup of organic surface films from the ocean surface by small droplets and sea-salt particles emitted into the atmosphere.

8.7.6
Modification of the Radiative Properties of Clouds

PROPERTIES OF CLOUDS

Clouds are one of the main agents in the global radiation balance and therefore in climate (see Section 5.2). They exert a dominant influence on the global and local albedo. Although the global surface albedo is usually between 0.05 and 0.20, the average global albedo is about 0.30 because of the high reflectivity of clouds in the shortwave spectrum.

Clouds (and also fogs) absorb (Bricard et al., 1971) and scatter visible radiation, and absorb, scatter, and emit infrared radiation. The magnitude of their effect on the radiation field depends on their type, composition (water or ice), vertical distribution and extent, and, especially for thin layers, on the number density and the size of the water droplets or ice crystals and on the albedo and emissivity of the surface below them. Optically thick clouds emit practically as blackbodies (the net radiative flux is different from zero only near the cloud boundaries). Computations show that they have an albedo relatively insensitive to optical thickness. Clouds and fogs of all types can be generated or enhanced by the introduction into the atmosphere of additional condensation and ice-forming nuclei due to direct or indirect man-made pollution (Conover, 1966; Twomey, Jacobowitz, and Howell, 1968; Conover, 1967).

These general considerations lead one naturally to consider the effect of possible changes in the optical properties of clouds or in their geometrical extent due to the increasing particle concentration of our atmosphere. The average optical properties of clouds as presently understood, including the possibility of absorption in the shortwave spectrum, are summarized in the SCEP Report (1970) and in its references. We shall deal here with the possible effects of the changes in optical properties as observed or as theoretically indicated.

CLOUDS AND SOLAR RADIATION

An increase, due to pollutants, in the droplet content of clouds will, for the same water content, lead to an *increase* of the visible albedo (in contrast to the views stated in the SCEP Report, 1970) and a decrease in transmissivity. This is because for droplets of radius 1 μ or more the optical thickness of a cloud of geometric depth h is, approximately, $2\pi Nr^2h$, where N is the concentration of drops per unit volume. The liquid water content W of the cloud per unit volume is $(4\pi/3)Nr^3$. Clearly, therefore, if h and W are held constant, the optical depth will *increase* with increas-

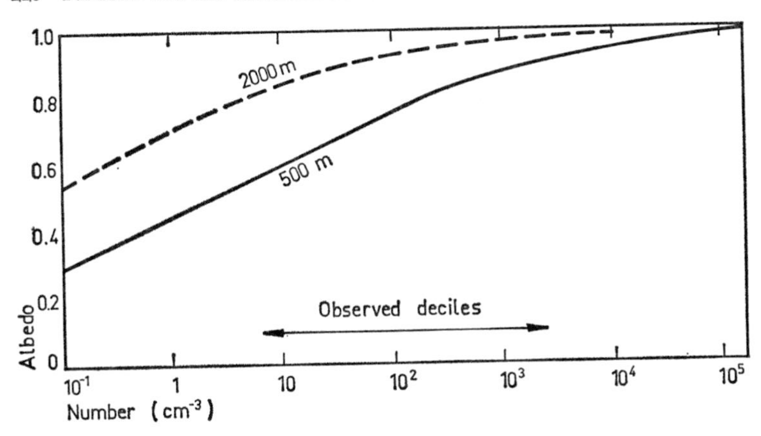

Figure 8.9 The visible albedo of a water cloud as a function of droplet content for thicknesses of 500 meters and 2000 meters.
Source: Twomey, 1971.

ing drop concentration. The optical depth τ will then be proportional to $N^{1/3}$. Thus while h and W will be fixed by large-scale circulation factors, the microphysics influences the optical properties through the $N^{1/3}$ term. The value taken by N is decided primarily by the concentration of cloud nuclei, that is, by particles.

With increasing optical thickness, a cloud reflects and absorbs more and transmits less. The effect can be quite sizable, as shown for instance in Figure 8.9 (Twomey, 1971). The albedo plotted is a computed average that takes into account the different directions of incident illumination falling on the spherical earth. The numerical method used includes all orders of multiple scattering and assumes no absorption. The range of values of particle numbers observed in measurements at a single station in southeastern Australia is indicated by arrows in the figure; eight-tenths of the observations there were within the range indicated by the double-ended arrow. The global range would obviously be wider.

On occasions the concentration of cloud nuclei may be so low that the clouds formed in the air mass will be relatively transparent optically. Conover (1966) has reported satellite observations in which a region of the North Pacific was apparently free from clouds everywhere except along the path of a ship where the additional nuclei "brightened" the clouds enough to make the clouds visible to the satellite camera.

CLOUDS AND INFRARED RADIATION

A sizable increase in albedo and emissivity together with a decrease in transmissivity with increasing optical depth also exists in the thermal infrared region of the spectrum as shown in Figure 8.10 (Yamamoto, Tanaka, and Kamitani, 1966; Yamamoto, Tanaka, and Asano, 1970; Kattawar and Plass, 1971; Korb and Zdunkowski, 1970). In this infrared region of the spectrum the main influence of clouds on radiation would be an increase in the downward thermal flux below the cloud leading to a decrease in the net flux and thus to an increase in the "greenhouse effect" (see Section 6.5.4).

It is difficult to speculate on the effect of thin clouds ($\tau < 1.0$) or of fogs on the atmospheric and surface heat balance because of their great variability in geometrical extent and optical thickness. In this case the visible albedo and infrared emissivity of the surface will have a major effect on the radiation fluxes and the radiative heating rates.

CONCLUSION

It is quite well established that particles can affect the droplet or particle concentrations in clouds. In turn the optical transmission,

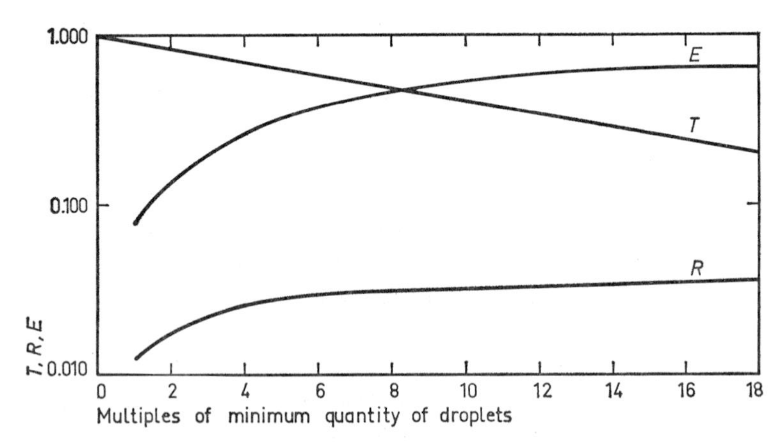

Figure 8.10 The transmissivity T, reflectivity R, and emissivity E of a water cloud at 240°K at 10 microns.
Source: Adapted from Yamamoto, Tanaka, and Kamitani, 1966, by Joseph, 1970.

reflection, and absorption are affected. Our knowledge of the various processess is still far from complete, especially where cirrus and other ice clouds are concerned and further research effort is called for. Since all particles do not participate directly in cloud formation, it would be desirable to monitor the concentration of those that do (cloud nuclei) and the concentration of those particles (ice nuclei) that can alter the microstructure of supercooled clouds and may be factors in cirrus formation.

Further effects of particles may be important, notably their possible effects on rain formation, and thereby indirectly on cloud amount and the global albedo (see Section 6.5.4). Mechanisms exist that tend to act in opposite directions, and again only more observation and research can clarify the relative importance of the various factors.

Pollution of clouddrops by solution of soluble gases and collection of particles is known to occur and is an important mechanism for the removal of particles. It may also affect the optical absorptive qualities of the clouddrops. The radiation absorbed by cloud layers is believed to be appreciable and would of course be very sensitive to the absorption of an individual drop. Because of multiple scattering, a small increase in absorption by the drops would produce a disproportionately large increase in the cloud absorption. Measurements are needed to clarify the nature and extent of absorption effects.

8.7.7
Recommendations

1. We recommend increased research on the refractive index (including absorption) of atmospheric particles in relation to their composition, origin, size, and shape and its change with increasing pollution for short- and longwave radiation. Also more data on particle growth with humidity are needed to understand the influence on the refractive index with relative humidity.
2. We recommend comprehensive comparative studies of the radiation fields for clean and polluted atmospheres in order better to identify the effects of short- and longwave radiation on the atmosphere and its modification with increasing pollution.
3. We recommend measurements of albedo and other radiative properties of clouds and fogs in unpolluted and polluted areas,

in conjunction with sufficiently complete measurements of cloud microstructure. These very important measurements require operation of instrumented aircraft.

4. We recommend studies of the effects of pollution on the refractive index of cloud droplets and ice crystals.

5. We recommend theoretical study of the integrated effects of pollution on radiative properties of clouds and fogs.

6. We recommend that comprehensive field studies directed toward the resolution of the question of the effect of particle concentration on frequency, type, and intensity of clouds and precipitation be designed and implemented.

7. We recommend that at suitable time intervals (about 2 years) the refractive index and the detailed size distribution of particles be determined at selected places in both clean and polluted air.

8. We recommend that objective methods for monitoring cloud cover by satellites or other means and for monitoring changes of cloud cover over large areas of the atmosphere be developed and implemented.

9. We recommend that at suitable intervals (for example, 5 years) the optical properties of the clouds (reflectivity, transmissivity, emissivity, and absorption) be determined in areas of increasing air pollution.

8.8
Gaseous Components

8.8.1
Introduction

Trace gases are gaseous components of the earth's atmosphere with great variation in time and space. The behavior of the various gases in the atmosphere differs widely. In relation to climate those atmospheric constituents that can influence directly or indirectly the radiation budget of the atmosphere are important. Although carbon dioxide and water vapor are not properly described as trace gases, it is convenient to consider their effects at the same time. In this section new results on the global trend and climatic effects of CO_2 obtained since the publication of the SCEP Report and the water vapor content of the middle and higher troposphere are discussed. The gaseous sulfur components and ammonia, which are basic substances for secondary particle formation, exert an indirect influence on the radiation field. In view of the importance of these gases as a source of atmospheric particles, their distribution in the troposphere is described in Section 8.3.3 rather than here. Several gases, particularly oxygen, methane, and carbon monoxide, have been widely discussed recently with respect to possible long-term trends. For oxygen a decrease has been suggested; for methane and carbon monoxide an increase. New information on these atmospheric components presented in this chapter reveals that they do not constitute a problem in the context of this study and that the statements made concerning them in the SCEP Report are still valid.

8.8.2 Carbon Dioxide

PREDICTION OF FUTURE CONCENTRATION

The prediction of future atmospheric concentrations of CO_2 depends on two factors: first, the amount of CO_2 that will be added to the atmosphere by the combustion of fossil fuels, and second, the fraction of this CO_2 that will remain airborne. The SCEP Report (1970) assumed that the world consumption of fossil fuels will continue to grow at about 4 percent per year until 1980. Thereafter, until 2000 A.D. the growth rate will drop to 3½ percent per year because of the likely expansion of nuclear sources of energy. The fraction of CO_2 from fossil fuel combustion remaining airborne has often been assumed to be the same in the future as it has been in the immediate past, that is, about 50 percent. The results of a recent analysis provide further confirmation for this assumption.

Figure 8.11 depicts the monthly values of CO_2 concentration at Mauna Loa, Hawaii (19° N). Across the bottom are given the yearly increases in concentration. Three fluctuations are evident. The first is a seasonal variation, decreasing during the summer growing season of the Northern Hemisphere. This seasonal variation appears at all clean air stations, the amplitude over Scandinavia (Bolin and Bischof, 1970) and northern Alaska (Kelley, 1969) is more than twice that found at Mauna Loa, which, in turn, is larger than that at the South Pole (Keeling, 1971). The second type of fluctuation is a long-term upward trend presumed to be

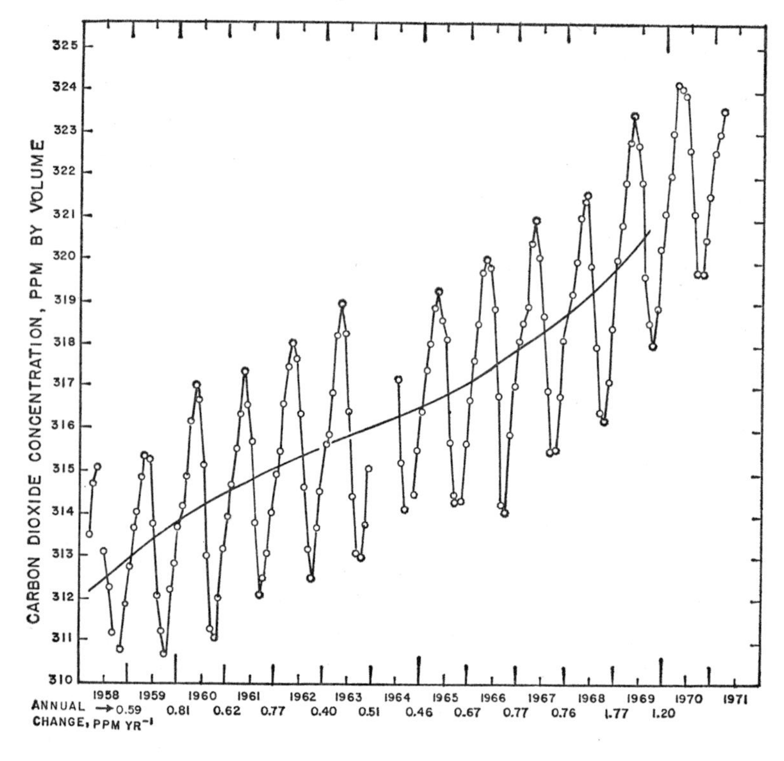

Figure 8.11 Mean monthly values of CO_2 concentration at Mauna Loa, Hawaii, for the period 1958 to 1971.
Sources: For 1958 to 1963, Pales and Keeling, 1965. For 1963 to 1971, Keeling and Bainbridge (unpublished), 1971. The source of the best-fit third-degree polynomial curve for the period 1958 to 1969 is Cotton, NOAA (National Oceanic and Atmospheric Administration, United States).

a consequence of the combustion of fossil fuels. From 1958 to 1968, the average annual increase amounted to 0.64 ppm/yr, while the average annual growth of fossil fuel CO_2 would have increased the global value by 1.24 ppm/yr. Thus about half of the fossil fuel CO_2 has remained airborne. The third type of fluctuation is the year-to-year change in the rate of increase.

Between 1962 and 1965, the annual growth rate dropped to 0.46 ppm/yr, and in the last two years the growth rate has increased to 1.48 ppm/yr. Data from Ship P (Wong, 1971), Ship C, and Niwat Ridge on the continental divide in the United States,

Scandinavia (Bolin, 1970), and Antarctica (Keeling, 1971), all confirm a growth rate of at least 1 ppm/yr after 1969, while the data that exist for the period before 1969 support the smaller growth rate of 0.6 to 0.7 ppm/yr between 1958 and 1968.

Machta (1971) has constructed a simple predictive model of the exchange of carbon dioxide between atmosphere, biosphere, and oceans (Figure 8.12), and using this it is possible to predict further concentration of atmospheric CO_2 given the future CO_2 production. The most crucial of the exchange coefficients is that between atmosphere and oceans. This has been found from the observed decrease of man-made $C^{14}O_2$ as reported by Telegadas (1971). All other exchange parameters are assumed.

The model contains a troposphere and stratosphere and a mixed and deep oceanic layer. The exchanges between each of these adjacent reservoirs obey first-order kinetics. The quantity λ is the fraction transferring between reservoirs. The biospheric uptake of CO_2 is given by the net primary biological production or the yearly fixation of organic carbon taken out of the air and the ocean. The CO_2 from fossil fuel combustion remaining airborne enhances the net primary production in proportion to the amount of extra CO_2 available to the plants and forests; a 10 percent increase in atmospheric CO_2 results in a 5 percent increase

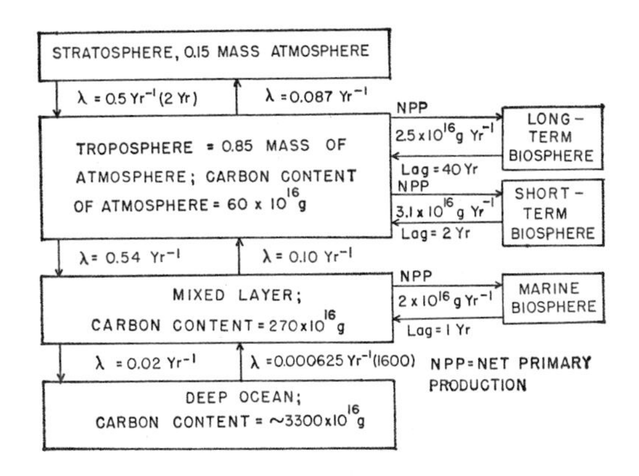

Figure 8.12 Simple predictive model of the exchange of CO_2 among the atmosphere, oceans, and biosphere.
Source: Machta, 1971.

in net primary productions, but this enhancement applies to only one-half of the land biosphere. The rest of the land biosphere is presumed to be limited by nutrient or water supply rather than by CO_2 supply and does not increase its growth in the presence of more CO_2.

The buffering action of additional CO_2 in the surface layers of the ocean limits the oceanic uptake of the fossil fuel CO_2. It is assumed, in effect, that a 1 percent increase in CO_2 in the mixed layer of the ocean water creates a 10 percent increase in the partial pressure of CO_2 in the water. The average residence time for CO_2 resulting from the model is about 2 years.

The prediction for atmospheric CO_2 between 1860 and 2000 A.D. is given in Figure 8.13. The computations assume a concentration of preindustrial plus fossil fuel CO_2 in 1958 of about 313 ppm. The year-to-year prediction for the period 1958 to 1970 has been enlarged in the figure. The calculations verify exactly over the whole time interval, but fail to reflect the departures from the average growth from year to year.

The prediction for the year 2000 A.D. is 375 ppm. The model indicates that about 50 percent of the fossil fuel CO_2 remains airborne, hence the favorable agreement with the prediction of 380 ppm given in the SCEP Report. The lower value, in fact, results from a reduction in the estimates of past releases of fossil fuel CO_2 given by Keeling (1971) and used as the basis upon which the growth in future years is based.

The model used in this prediction is the simplest that can provide useful results. Recomputation of the predictions for the year 2000 A.D. using large variations in parameters but still fitting the man-made atmospheric $C^{14}O_2$ values provides forecasts between about 365 and 385 ppm. In addition to uncertainties in model formulation and in the choice of model parameters, errors in the inventories of man-made $C^{14}O_2$ or in the forecasts of fossil fuel emissions can create departures greater than 10 ppm from the predicted value of 375 ppm in the year 2000 A.D.

Perhaps the most serious uncertainty in a forecast of future atmospheric CO_2 stems from the inability to explain why there are year-to-year changes in the annual growth rate beyond that expected from the changes in fossil fuel production. If the bio-

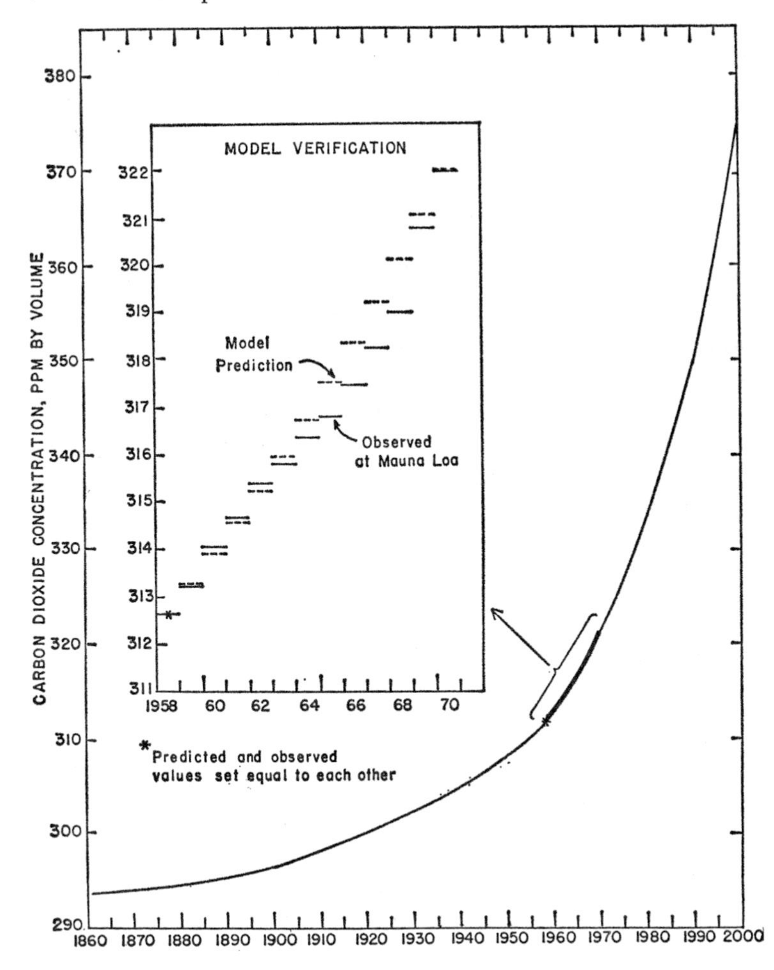

Figure 8.13 Model calculation of atmospheric CO_2 from combustion of fossil fuels.
Source: Machta, 1971.

sphere or oceans are significantly altering their capacity to take up atmospheric CO_2, then it is necessary to understand and predict these characteristics as well as their average behavior over the past 10 years.

CLIMATIC EFFECTS OF CHANGING CONCENTRATION

As early as in 1863, J. Tyndall expressed the opinion that changes in the CO_2 concentration in the atmospheric air could influence the climate near the earth's surface. The gas has a strong absorption band around 15 μ and in this band emits longwave infrared radiation downward to the ground and upward to space. The downward radiation will increase with increasing CO_2 concentration, diminishing the heat loss of the earth's surface. The question was further studied by Arrhenius (1896), Plass (1956), Kaplan (1960), Möller (1963), Manabe and Wetherald (1967), Rasool and Schneider (1971), and others. Manabe and Wetherald's results still appear to be the most reliable ones using a global average model (see also Manabe, 1971). They found that a doubling of the carbon dioxide concentration will increase the average surface temperature by 2°C; an increase of atmospheric carbon dioxide from 320 to 375 ppm will warm the surface layer by about 0.5°C. The model calculations also indicate that stratospheric temperatures will decrease.

There are other processes that are important in the Manabe and Wetherald calculations. It is probable that, if the troposphere warms, the relative rather than the absolute humidity would remain constant and the total amount of moisture in the atmosphere would increase with time (Möller, 1963). The increasing total amount of water vapor would effect a further shielding of the infrared emission from the ground and act as a positive feedback mechanism (see Sections 6.4.2 and 6.8.2). This increased tropospheric heating, as well as that due to increased solar radiation absorption by more carbon dioxide and water vapor has been taken into account in these calculations.

Manabe (private communication) has recently computed the distribution of temperature changes resulting from the doubling of CO_2 concentration in air by use of highly simplified, three-dimensional model of the general circulation. His model has a limited computational domain and a highly idealized topography. It has no heat transport by ocean currents (the ocean is treated as a wet "swamp" without any heat capacity). Nevertheless, the results from his numerical experimentation yield some indication of how the increase in CO_2 concentration can affect the distribution of temperature in the atmosphere. His results show somewhat larger general warming of troposphere than that expected from the global average model—owing to the recession of snow cover, another positive feedback—and they indicate a marked geographical variability. Particularly, the surface air temperature in high latitudes exhibits a twofold greater warming than the global average, due to a positive feedback mechanism of the ice and to the thermal stability of the lower troposphere which limits convective heating to the lowest layer (see Section 5.2.2).

One of the important factors that have not been taken into account in any of the models is a possible change in cloudiness. An increase in the amount of low cloud over the whole globe by only 2.4 percent could lower the average surface temperature by about 2°C, negating the carbon dioxide warming. An increase of only 0.6 percent in low cloud cover could cool the lower atmosphere by 0.5°C. It is almost certain that meteorologists will be unable to detect, let alone predict, global cloud cover changes as small as 0.6 percent in the foreseeable future. As far as can be seen, there is no simple coupling between temperature and cloudiness (see Section 6.5.4), and it should not be forgotten that many factors controlling radiation, such as cloudiness and the global albedo, may be undergoing undetected secular trends due to natural or perhaps human activities.

CONCLUSION

Doubling of the CO_2 concentration could effect an increase of the temperature near the surface by about 2°C, while the increase from 320 to 370 ppm expected by the year 2000 A.D. could result in an increase of the temperature by 0.5°C. The 2°C change would constitute a modification of the climate which could trigger other warming mechanisms and possibly lead to irreversible effects (see, for example, Sections 6.7 and 7.3.1). Variations of other meteorological parameters may moderate the warming; for instance, an increase of low cloud coverage of only 0.6 percent could decrease the temperature by the same amount, 0.5°C.

8.8.3
Water Vapor

As discussed earlier in Chapter 5, atmospheric water, in both vapor and cloud form, is a significant parameter in the understanding of climate and climatic changes because it plays an important role in the atmospheric energy cycle—through evaporation and condensation processes—and in the absorption, emission, and reflection of radiation.

Water vapor is transported upwards (from its oceanic source) through the mid-troposphere, where, as a result of the low temperatures at these levels, condensation occurs, forming clouds. In general, the liquid water precipitates back to the earth's surface; thus the water vapor mixing ratio decreases with height by approximately one-third for every 2 kilometers.

Usually, except for example when it is transformed to clouds near the tropical tropopause, water vapor is the major atmospheric gas acting to cool the troposphere radiatively. Thus, data concerning its distribution are necessary for a correct assessment and understanding of its radiative effects. The seasonal and latitudinal distribution of water vapor is reasonably well known up to heights of about 6.5 kilometers, but there are few routine measurements above this level. The total amount of water vapor in a unit column of air, expressed as precipitable water vapor, decreases from a maximum of about 50 millimeters in equatorial regions to a minimum of about 5 millimeters over the arctic area and 1.5 millimeters over the antarctic area. In general, the water vapor concentration near the surface follows the temperature distribution and is higher during summer than winter. In tropical regions of both hemispheres the observations show that the relative humidity is about 60 to 75 percent near the surface, decreasing to about 50 to 60 percent in mid-troposphere, but in subtropical latitudes the minimum relative humidity is approximately 30 percent at a height of about 5 kilometers. At present, there is no evidence that the tropospheric water vapor concentration is changing either naturally or as a result of the varied activities of man.

In Section 9.4 we shall see the importance of the water vapor content of the stratosphere. Water vapor probably reaches the stratosphere through transport upwards in the deep convective circulation in the tropics (although a small amount of water vapor may enter the stratosphere through thunderstorm clouds and circulation process through the mid-latitude tropopause). In order to undertand better its distribution and transport at these levels, it is necessary to extend routine water vapor measurements with relative accuracy of about 5 to 10 percent. At the same time it is essential that sufficient observations, both in number and accuracy, be made in the stratosphere to be able to determine the absolute magnitude of the water vapor concentration in this region as well as its latitudinal and temporal variations.

8.8.4
Nitrogen Dioxide and Reactive Hydrocarbons

The atmospheric phenomenon commonly known as "photochemical smog" first came to attention as a major problem in the city of Los Angeles in the United States, but it has increasingly become of worldwide concern, particularly in urban communities. Sunlight and the presence of reactive hydrocarbons and oxides of nitrogen formed by combustion mechanisms (of which the internal combustion engine is the prime example) are the agents for its formation.

No definite and comprehensive reaction scheme has been derived from which a complete understanding of the smog-formation process can be achieved. The report, "Cleaning our Environment—The Chemical Basis for Action," of the American Chemical Society (1969) describes the general, qualitative, features of the process as beginning

> when nitrogen dioxide absorbs ultraviolet light from the sun and is broken down to nitric oxide and atomic oxygen, atomic oxygen reacts with molecular oxygen to form ozone, which reacts in turn with nitric oxide to form nitrogen dioxide and molecular oxygen. Atomic oxygen also reacts with reactive hydrocarbons to form chemical species called radicals. These radicals take part in a series of reactions involving the formation of more radicals, which react with molecular oxygen, hydrocarbons, and nitric oxide. Nitrogen dioxide is regenerated, the nitric oxide eventually disappears, and ozone begins to accumulate and react with hydrocarbons. Second-

ary pollutants are formed, including formaldehyde and other aldehydes, ketones, and peroxyacyl nitrates (PAN).

Although the chemical constituents of the smog are almost continuously present in the atmosphere, significant levels are not an everyday occurrence at any location. These require particular conditions of air concentrations of the precursors, solar radiation, atmospheric dispersion, and residence time over the source region of the primary reactants. The most important effects of the photochemical smog on climatic elements are in the reduction of visibility and on those elements in which particles, per se, play a role, that is, radiation field effects and precipitation mechanisms. Effects on visibility are local in the case of man-made photochemical smogs and regional in the case of "blue-haze" phenomena resulting from photooxidation of terpenes emitted by vegetation. Considering the biosphere as a whole, it is evident from Table 8.1 that the emission and formation of hydrocarbons and oxides of nitrogen into the atmosphere from "natural" sources probably exceed to a considerable extent those associated with man-made (primarily urban) emissions. Together they comprise an important contribution to the total particulate loading of the atmosphere.

8.8.5
Methane and Carbon Monoxide

To the best of our knowledge most atmospheric CH_4 is produced by microbiological activity in soil and swamps under anaerobic conditions, and it is fairly certain that most of it is also destroyed again by microbiological action under aerobic conditions on the surface. The stratosphere constitutes only a very minor sink. The estimated annual natural production is so large that any anthropogenic sources constitute minor fractions. For this reason, and because CH_4 has no direct effects on the climate or the biosphere, it is considered to be of no importance for this report.

Very recent research indicates that CO is produced by the oceans in considerable quantities and destroyed at a rapid rate in the stratosphere and at the earth's surface. We must therefore assume that CO has always been a natural constituent of our atmosphere (Junge, Seiler, and Warneck, 1971; Inman et al., 1971).

With an average lifetime of CO of about 1 year, the present concentration of about 0.1 ppm by volume represents a steady-state situation, so that a continuous accumulation with time, due to production of CO primarily by automobile exhaust, is not possible. The steady-state concentration of CO, however, must have increased as a result of man's activity by factors varying from 2 to 20 according to present estimates. There is a possibility that increasing CO production may have a slight influence on the stratospheric ozone level, but apart from this no adverse effects of CO on the climate can be expected.

8.8.6
Oxygen

The SCEP Report (1970) considered the problem of a decrease of atmospheric oxygen which could conceivably be brought about by combustion and/or through a possible reduction of vegetable growth, particularly in the oceans, by pesticides, herbicides, or pollutant poisons because vegetation is responsible for generation of oxygen during the photosynthetic fixation of the carbon of atmospheric CO_2. The conclusion drawn was that measurements of oxygen, the most recent by Machta and Hughes (1970), showed no detectable change in the atmospheric oxygen content and that consumption of oxygen by burning all the "recoverable reserves of fossil fuel" would deplete the oxygen content of the atmosphere by only 0.7 percent. Davitaya (1971) has again raised the question, apparently unaware of Machta and Hughes's measurement. He calculates that fossil fuel consumption during the last 50 years could have reduced the atmospheric oxygen content by about 0.02 percent. The accuracy is such that the measurements do not conclusively disprove Davitaya's suggestion; however, Davitaya also computes the amount of oxygen that would be consumed by combution of all the carbon which exists in oxidizable form ("all fossil fuel"). He estimates the amount of O_2 would be 3×10^{16} kg, which could account for about 10 percent of the atmospheric oxygen if there were no replacement or regeneration. This level of recovery of fuel is most unlikely to be reached, and Davitaya's postulated exponential increase of consumption rate is even more unlikely. There seems to be no reason to revise the conclusions of the SCEP Report that oxygen decrease is not at present a problem, but that monitoring by careful measurements at intervals of about 10 years is advisable.

8.8.7
Recommendations

CARBON DIOXIDE

1. We recommend that stations be established in clean air locations to monitor the trends in atmospheric carbon dioxide. A minimum of ten stations is required to sample different geographic and climatic regimes over the earth and to allow for possible later loss of some stations due to local influence. Aircraft collections form a valuable part of the network.
2. We recommend extensive study and survey in both time and space of the partial pressure of CO_2 in the surface layers of the oceans.
3. We recommend that research in the mechanisms of CO_2 uptake by the ocean and biosphere be undertaken in the hope that it will lead to an understanding of the observed changes in atmospheric CO_2. Attempts to survey the CO_2 content of both ocean and biomass should be made from time to time, in the light of improvements in our knowledge of these processes.
4. We recommend that continued computation of past records and updating of forecasts of the global consumption of fossil fuels in combustion be maintained.
5. We recommend study of the combined effect on the surface air temperature of increasing CO_2 content including the effect of changing cloud cover and snow cover. This should have high priority during the development of climatic models.

WATER VAPOR

6. We recommend that high priority be given to extension of the routine water vapor measurements in the upper troposphere. Relative accuracy of not less than ± 10 percent is required.
7. We recommend that more, and more accurate, observations be made in the stratosphere to determine the absolute magnitude of the water vapor concentration in this region as well as its latitudinal and temporal variations. Measurements to an accuracy of 1 ppm are required.

OTHER GASES

8. We recommend monitoring of SO_2, H_2S, and NH_3, concentrations in unpolluted areas, with an accuracy of ± 10 percent. Discontinuous measurement at about 10 base-line stations and 100 regional stations is required in order to improve the understanding of the life history of these gases and of the particle formation process in which they are involved.

OXYGEN

9. We recommend monitoring of O_2 by careful measurements in intervals of about 10 years at not more than 5 base-line stations. Accuracy is required to at least the fifth significant figure.

8.9
Cirrus Clouds
8.9.1
Climatic Importance

The addition of water vapor to the troposphere by jet aircraft could, conceivably, increase the cirrus coverage and/or thicken existing cirrus clouds. Two effects from such cirrus clouds can be expected. The effect about which we are most certain would be the production of a change in radiation balance. During daylight hours, the cirrus would reflect solar radiation, and the associated loss of solar heating at the earth's surface might not be offset by the downward longwave radiation from the cloud (see Section 6.5.4). Hence, during the day, an increased cirrus cloudiness might result in a small drop in surface air temperatures, depending on the optical properties of the cirrus (about which we know very little). At night, an enhanced cirrus coverage would raise surface temperatures, but the amount would depend not only on the height (that is, temperature) of the cloud but also on its microstructure, as is clearly indicated by measurements reported by Kondratiev (1965). We know too little about the microstructure of cirrus clouds to be certain about the magnitude of these effects. Manabe and Wetherald (1967) have carried out a numerical integration concerning the effects of an enhanced cirrus coverage. Their results show that an increase in cirrus cloud amount could lead to a rise of a few tenths of °C in surface air temperatures.

However, the magnitude of the calculated effect depends on the optical properties adopted for the cirrus. In nature, these optical properties appear to vary considerably. Measurements by Kuhn and Weickmann (1969) showed a variation in infrared transmission of between 95 and 50 percent, the latter in thick cirrus.

The second expected effect of an enhanced cirrus cover would be that ice crystals falling out from cirrus might nucleate supercooled low clouds and thus initiate precipitation sooner than it would otherwise occur.

8.9.2
Water Vapor Emission by Jet Aircraft

According to information available to us (Beckwith, private communication), the estimated consumption of kerosene by jet aircraft in the United States (refueling for both national and international flights) in 1971 will reach about 5×10^{13} g/yr. We do not have at this time similar figures for other countries, and it is virtually impossible to obtain the consumption by military planes. We shall assume, tentatively, that the world consumption, civilian as well as military, is four times the U.S. civilian consumption, or about 2×10^{14} g/yr. On this basis, the water vapor added would amount, for the world as a whole, to about 2.5×10^{14} g H_2O/yr. In middle latitudes in summer, the jets fly predominately in the upper troposphere, whereas in winter the same cruising altitude is in the lower stratosphere. We shall assume that the appropriate residence time in the upper troposphere is 30 days, while that for the lower stratosphere is 120 days (SCEP, 1970) with the result that our estimates for the added water vapor are 2×10^{13} g H_2O in the upper troposphere and 8×10^{13} g H_2O in the lower troposphere.

We assume, as an approximation, that the added water vapor mixes uniformly in an air layer 200 millibars deep. Most of the flights take place between about 30° and 60° N, and we assume that the added water vapor will not move out of this latitude band. We arrive at the following figures for increase in water vapor mixing ratios:

Upper troposphere (30° to 60° N)	10^{-4} g H_2O/kg air
Lower stratosphere	4×10^{-4} g H_2O/kg air

Locally and temporarily, relatively large deviations from these values would be expected.

The mean annual mixing ratio of water vapor at heights 12 to 14 kilometers, where most of the jets fly at present, is (Gutnick, 1965) between 1 and 2×10^{-2} g H_2O/kg air. We thus see that the "equilibrium additions" of water vapor are nearly two orders of magnitude smaller than the mean values. We estimate that the natural variations about the mean are greater than one order of magnitude. It appears from a recent study by Van Valin (1971) that at heights between 12 to 14 km in low latitudes (15° N) the saturation mixing ratio over ice is about 0.12 to 0.5 g H_2O/kg air and the parallel quantity with respect to water is about 0.18 to 0.8 g H_2O/kg air. In high latitudes (75° N) in winter, when the lower stratosphere is very cold, the corresponding figures are lower by about one order of magnitude.

Commercial aircraft flight may increase by a factor as much as 3 to 6 in the years 1985 to 1990. Thus, the large natural range of variations in mixing ratio is likely to offer occasions in the not-so-distant future when the emissions by the increasing numbers of jets will create conditions of saturation with respect to ice and, possibly, with respect to water. Such a possibility does not appear to be at all remote, in areas where the jet air traffic is dense, as, for instance, over the United States and perhaps western Europe. Hence, in the 1980s the likelihood of cirrus formation or that of thickening of cirrus clouds seems to be rather definite.

8.9.3
Observed Changes of Cirrus Cloud Cover

Machta and Carpenter (1971) report a study of secular change in the amount of high cloud. Figures 8.14 and 8.15 show the history, by year, of average high cloud cover when no low or middle clouds were present. The dashed lines in the lower right of each figure should be read from the scale on the right and denote roughly the increase in jet traffic over the United States since 1958. The horizontal lines are the mean values of cloudiness over the indicated period.

It is evident that since 1965 all stations show an increase in high cloudiness over the 1949 to 1964 average. Denver has almost twice as much, but the other stations in Figure 8.15 contain only

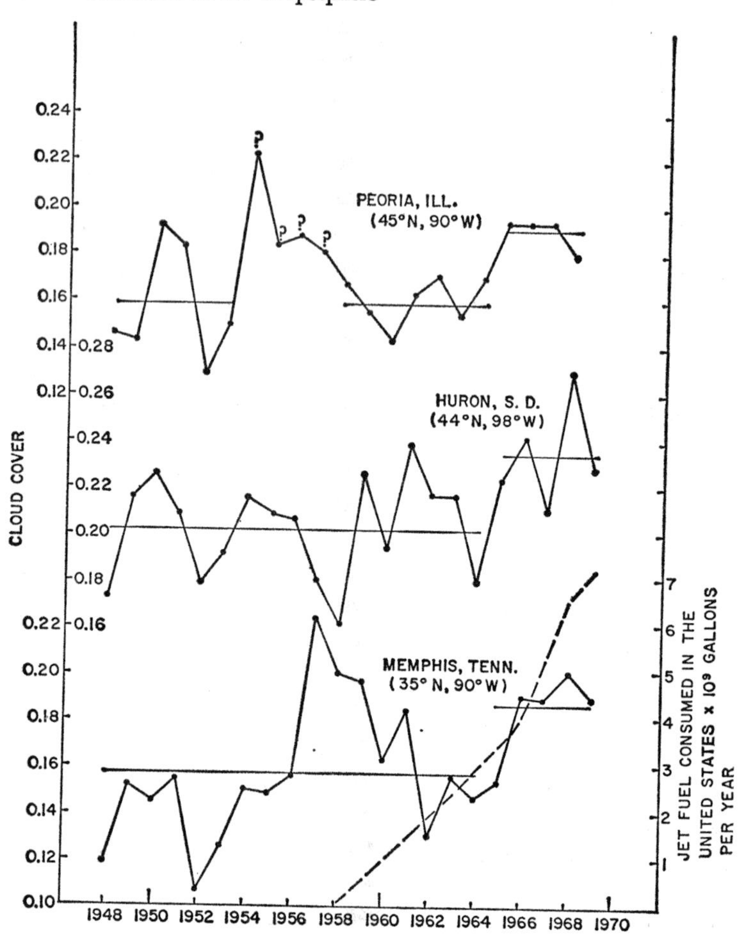

Figure 8.14 Average annual high cloud cover with zero low or middle clouds for Peoria, Illinois, United States, Huron, South Dakota, United States, and Memphis, Tennessee, United States.
Source: Machta and Carpenter, 1971.

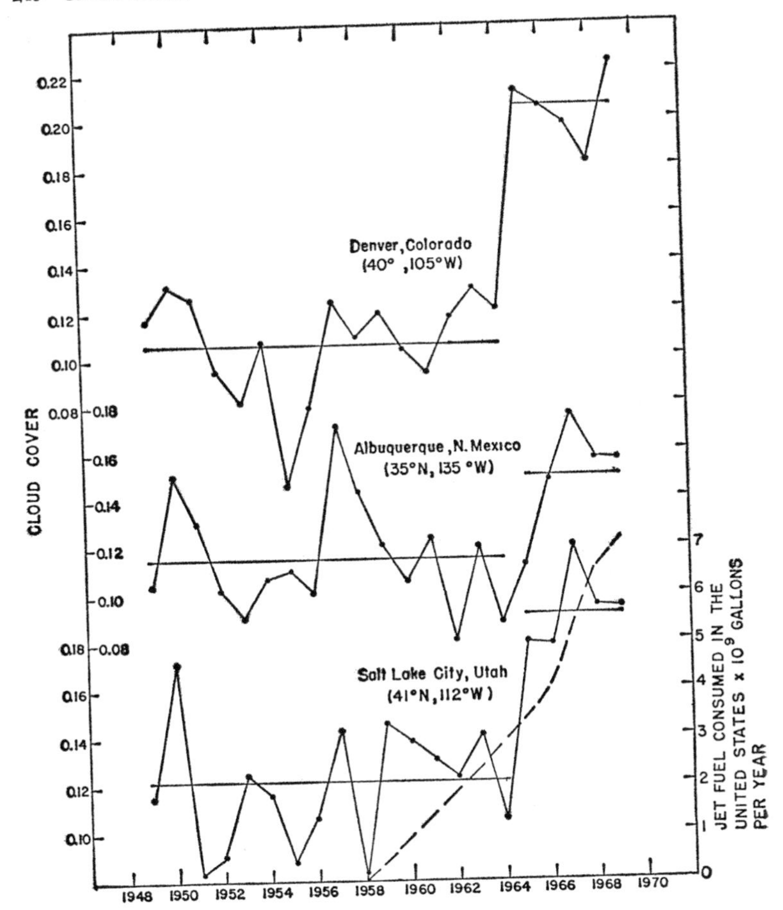

Figure 8.15 Average annual high cloud cover with zero low or middle clouds for Denver, Colorado, United States, Albuquerque, New Mexico, United States, and Salt Lake City, Utah, United States.
Source: Machta and Carpenter, 1971.

17 percent more. Thus, it appears safe to conclude that high cloudiness has increased over the United States since 1965.

The coincidence of more high clouds with increased commercial jet flights suggests a causal relation, but two significant bits of information suggest caution. First, there appears to be nothing unique about 1965 in aircraft operations, when the rise in high cloud cover first appeared. Second, an analysis of middle cloud cover during times with no low clouds indicates a marked decrease at five of the six stations of Figures 8.14 and 8.15 in the 1950s. Insofar as can be ascertained, these changes in middle cloud cover are of natural origin. Therefore, it may be argued, natural causes may also have produced more high clouds after 1964.

8.9.4 Recommendations

1. We recommend that high priority be given to efforts to determine the humidity at which cirrus cloud forms. In particular, the question of whether cirrus forms through sublimation or by means of the liquid phase should be answered.
2. We recommend that information collected on the fundamental physical properties of cirrus clouds. These include water content, particle concentration, and distribution of sizes and shapes of crystals.
3. We recommend that the optical properties of cirrus clouds, in both solar and infrared radiation be investigated.
4. We recommend that high priority be given to monitoring trends, if any, in cirrus cloudiness and characteristics. For this purpose objective methods are needed in order to distinguish between cirrus and lower clouds.

8.10 Some Implications for Implementation of the Recommendations

It is possible to make some comments that apply generally to the recommendations concerned with measurement of the composition of the troposphere. They call for a variety of physical and chemical measurements of varying degrees of difficulty. A very few of them, for example, those employing a sun photometer can be performed by relatively unskilled personnel, and many can be performed by a technician trained in the specific task. All, even the simplest, call for rigorous attention to procedures, and all call for some degree of standardization and central inspection. A few, such as the determination of refractive index of the material of atmospheric particles, are still research problems.

The short atmospheric life and the geographical variation of the concentration of much of the material of interest necessitate establishment of many stations—the SCEP recommendation of about 100 is repeated. The requirement for planning, control, and standardization to maintain the regularity and particularly the accuracy of a network of observations stretching across national frontiers and covering international waters thus arises. We understand that a United Nations Specialized Agency is prepared to accept this responsibility for some of the measurements we recommend. As noted, many of the recommendations are similar to those made by SCEP (1970), and the SCEP estimates of the cost of implementation are still a reasonable guide.

Many of the recommendations for studies involve the use of aircraft, and for this reason are expensive in the context of the resources generally allocated to scientific research that has neither commercial nor military implications. The effort called for is a very small proportion of the current world total of aircraft hours. Some of it might be accommodated by more efficient use of aircraft facilities already available to atmospheric scientists, but it would be counterproductive to call for any substantial shift in the priorities already established for use of these facilities. Implementation of the recommendations implies a substantial increase in the aircraft facilities available for research and monitoring in the atmosphere.

The recommendations concerning the CO_2 content of the atmosphere are in general similar to those made by SCEP, but the distinction between research necessary to understand and monitoring necessary to provide warning is not made, and the recommendations concerning ocean measurements are far less specific. There is further support in these recommendations for the thesis developed in many contexts in SCEP and SMIC that many oceanic and atmospheric problems are inseparable, with the present level of understanding of the oceanic aspect lagging behind that of the atmospheric aspect.

References

Additional references for Chapter 8 are on page 294.

Atwater, M. A., 1970. Planetary albedo change due to aerosols, *Science, 1:* 64–66.

Beilke, S., 1970. Laboratory investigation on washout of trace gases, *Precipitation Scavenging,* Proceedings of symposium held at Richland, Washington, June 2–4, 1970, sponsored by Pacific Northwest Laboratory, Battelle Memorial Institute, and the Fallout Studies Branch, Division of Biology and Medicine, U.S. Atomic Energy Commission, AEC Symposium Series No. 22 (Washington, D.C.: U.S. Atomic Energy Commission), *261,* p. 1128.

Beilke, S., and Georgii, H. W., 1968. Investigation on the incorporation of sulfur-dioxide into fog and rain droplets, *Tellus, 20:* 435–442.

Bigg, E. K., and Stevenson, C. M., 1970. Comparison of concentrations of the nuclei in different parts of the world, *Journal des Recherches Atmosphériques, 4:* 41–58.

Bolin, B., 1970. Carbon cycle, *Scientific American, 223*(3): 124–132.

Bolin, B., and Bischof, W., 1970. Variations of the carbon dioxide content of the atmosphere in the Northern Hemisphere, *Tellus,* 22.

Bricard, J., Billard, F., Cabane, M., and Madelaine, G., 1968. Formation and evolution of nuclei of condensation that appear in the air initially free of aerosol, *Journal of Geophysical Research, 73:* 4487–4496.

Bricard, J., Cabane, M., Madelaine, G., and Vigla, D., 1971. Formation and properties of ultrafine particles and small ions conditioned by gaseous impurities of the air (unpublished).

Büchen, M., and Georgii, H., 1971. Ein Beitrag zum atmosphärischen Schwefelhaushalt über dem Atlantik (unpublished).

Budyko, M. I., 1969. The effect of solar radiation variations on the climate of the earth, *Tellus, 21:* 611–619.

Bullrich, K., 1964. Scattered radiation in the atmosphere and the natural aerosol, *Advances in Geophysics, 10:* 101–260.

Charlson, R. J., Ahlquist, N. C., and Horvath, H., 1968. On the generality of correlation of atmospheric aerosol mass concentration and light scatter, *Atmospheric Environment, 2:* 455–464.

Cobb, W. E., and Wells, H. J., 1970. The electrical conductivity of oceanic air and its correlation to global atmospheric pollution, *Journal of Atmospheric Sciences, 27,* 814.

Conover, J. H., 1966. Anomalous cloud lines, *Journal of Atmospheric Sciences, 23:* 778–785.

Conover, J. H., 1967. Reply to comments, *Journal of Atmospheric Sciences, 25:* 355.

Cox, R. A., and Penkett, S. A., 1971. Photo-oxidation of sulfur dioxide in air (unpublished).

Danielson, R. E., Moore, D. R., and van de Hulst, H. C., 1968. The transfer of visible radiation through clouds, *Journal of Atmospheric Sciences, 26:* 1078–1087.

Davitaya, F. F., 1971. History of the atmosphere and dynamics of its gaseous content, *Meteorologiya i Hidrologiya,* No. 2:21–28 (in Russian).

deBary, E., and Junge, C., 1963. Distribution of sulfur and chlorine over Europe, *Tellus, 15:* 370.

Deirmendjian, D., 1969. *Electromagnetic Scattering on Spherical Polydispersions* (New York: Elsevier), pp. 290.

Eiden, R., 1966. The elliptical polarization of light scattered by a volume of atmospheric air, *Applied Optics, 5:* 569.

Elterman, L., 1964. Rayleigh and extinction coefficients to 50 km for the region .27 μ to .55μ, *Applied Optics, 3:* 1139–1147.

Engelmann, R. I., and Slinn, W. G., 1970. *Precipitation Scavenging,* Proceedings of symposium held at Richland, Washington, June 2–4, 1970, sponsored by Pacific Northwest Laboratory, Battelle Memorial Institute, and the Fallout Studies Branch, Division of Biology and Medicine, U.S. Atomic Energy Commission, AEC Symposium Series No. 22 (Washington, D.C.: U.S. Atomic Energy Commission), *261.*

Eriksson, E., 1959. The yearly circulation of chloride and sulfur in nature: Meterological, geochemical and pedeological implications, I, *Tellus, 11:* 375–403.

Fischer, K. 1970. Measurements of absorption of visible radiation by aerosol particles, *Atmospheric Physics, 43:* 244–254.

Flowers, E. C., McCormick, R. A., Kurfis, K. R., 1969. Atmospheric turbidity over the United States 1961–1966, *Journal of Applied Meteorology, 8:* 955–962.

Georgii, H. W., 1965. Untersuchungen über Ausregnen und Auswaschen atmosphärischer Spurenstoffe, *Berichte des Deutschen Wetterdienstes,* No. 106.

Georgii, H. W., 1967. Die Verteilung von Spurengasen in reiner Luft, *Experimentia Supplementum, 13:* 14.

Georgii, H. W., 1970. Contribution to the atmospheric sulfur budget, *Journal of Geophysical Research, 75:* 2365.

Goldberg, E., 1971. Atmospheric dust—the sedimentary cycle and man, *Geophysics: Earth Sciences* (in press).

Greenfield, S. H., 1957. Rain scavenging of radioactive particulate matter from the atmosphere, *Journal of Meteorology, 14:* 115.

Gutnick, M. 1965. Atmospheric water vapor, *Handbook of Geophysics and Space Environments,* edited by S. L. Valley, United States Air Force, Cambridge Research Laboratories (New York: McGraw-Hill Book Co.), pp. 3-34–3-37.

Hänel, G., 1968. The real part of the mean complex refractive index and the mean density of samples of atmosphere aerosol particles, *Tellus, 20:* 371–379.

Hidy, G. M., and Brock, J. R., 1970. An assessment of the global resources of tropospheric aerosols (Washington, D.C.: Second International Clean Air Congress, December 1970), Paper ME-26-A.

Hodge, P. W., 1971. A large scale decrease in the clear air transmission of the atmosphere 1.7 km above Los Angeles, *Project Astra* (Seattle: Astronomy Department, University of Washington).

Ikebe, Y., and Kawano, H., 1970. Dependence of the effective attachment coefficient of small ions upon the size of condensation nuclei, *Pure and Applied Geophysics, 83:* 120–130.

Itagaki, K., and Koenuma, S., 1962. Altitude distribution of fallout contained in rain and snow, *Journal of Geophysical Research, 67*(10): 3927–3933.

Joseph, J. H., 1970. Thermal radiation fluxes through optically thin clouds, *Israeli Journal of Earth Sciences, 19:* 51–67.

Joseph, J. H., 1971. Thermal radiation fluxes near the sea surface in the presence of marine haze, *Israeli Journal of Earth Sciences, 20:* 7–12.

Junge, C., 1955. The size distribution and aging of natural aerosols as determined from electrical and optical data on the atmosphere, *Journal of Meteorology, 12:* 13–25.

Junge, C., 1963. *Atmospheric Chemistry and Radioactivity* (New York: Academic Press).

Junge, C., and Jaenicke, R., 1971. New results in background aerosol studies from the Atlantic Expedition of the R. V. "Meteor," Spring 1969, *Journal of Aerosol Sciences* (in press).

Junge, C., Seiler, W., and Warneck, P., 1971. The atmospheric ^{12}CO and ^{14}CO budget, *Journal of Geophysical Research, 76:* 2866.

Kaplan, L. D., 1960. The influence of carbon dioxide variations on the atmospheric heat balance, *Tellus, 12:* 204–208.

Kasten, F., 1968. Der Einfluss der Aerosol-Grössenverteilung und ihre Änderung mit der relativen Feuchte auf die Sichtseite, *Beiträge Phys. Atm., 41:* 33–51.

Kattawar, G. W., and Plass, G. N., 1971. Influence of aerosol clouds and molecular absorption on atmospheric emission, *Journal of Geophysical Research, 76:* 3437–3444.

Kitaoka, T., 1959. On the "Trübungsfaktor" of the atmosphere over East Asia observed with silver-disc pyrheliometers, *Geophysical Magazine, 29:* 173–228.

Köenig, L. R., 1963. The glaciating behavior of small cumulonimbus clouds, *Journal of Meteorology, 20:* 29–47.

Kondratiev, K. Ya. 1965. *Radiative Heat Exchange in the Atmosphere* (London: Pergamon Press), p. 220.

Korb, G., and Zdunkowski, W., 1970. Distribution of radiative energy in ground fog, *Tellus, 22*(3): 298–320.

Kuhn, P. M., and Weickmann, H. K., 1969. High altitude radiometric measurements of cirrus, *Journal of Applied Meteorology, 8:* 147–154.

List, Robert J., 1963. *Smithsonian Meteorological Tables* (Washington, D.C.: The Smithsonian Institution), Tables 108, 109.

Machta, L., 1971. The role of the oceans and the biosphere in the carbon dioxide cycle (Gothenburg, Sweden: Nobel Symposium 20, August 16–20).

Machta, L., and Carpenter, T., 1971. Trends in high cloudiness at Denver and Salt Lake City, *Man's Impact on the Climate,* edited by W. H. Matthews, W. W. Kellogg, and G. D. Robinson (Cambridge, Massachusetts: The M.I.T. Press), pp. 410–415.

Machta, L., and Hughes, E., 1970. Atmospheric oxygen in 1967 to 1970, *Science, 168:* 1582–1584.

Manabe, S., 1971. Estimate of future change of climate due to the increase of carbon dioxide concentration in the air, *Man's Impact on the Climate,* edited by W. H. Matthews, W. W. Kellogg, and G. D. Robinson (Cambridge, Massachusetts: The M.I.T. Press), pp. 249–264.

Manabe, S., and Wetherald, R. T., 1967. Thermal equilibrium of the atmosphere with a given distribution of relative humidity, *Journal of Atmospheric Sciences, 24:* 241–259.

Martell, Edward A., 1971. Residence times and other factors influencing pollution of the upper atmosphere, *Man's Impact on the Climate,* edited by W. H. Matthews, W. W. Kellogg, and G. D. Robinson (Cambridge, Massachusetts: The M.I.T. Press), pp. 421–428.

Misaki, M., Ohtagako, M., and Kamazawa, I., 1971. Mobility spectrometry of the atmospheric ions in relation to atmospheric pollution (unpublished).

Mitchell, J. M., 1970. A preliminary evaluation of atmospheric pollution as a cause of the global temperature fluctuation of the past century, *Global Effects of Environmental Pollution,* edited by S. F. Singer (Dordrecht, Holland: Reidel Publishing Company; New York: Springer-Verlag), pp. 139–155.

Möller, F., 1963. On the influence of changes in CO_2 concentration in air on the radiation balance of the earth's surface and on the climate, *Journal of Geophysical Research, 68:* 3877–3886.

Mossop, S. C., Ruskin, R. E., and Heffernan, K. J., 1968. Glaciation of a cumulus at approximately –4°C, *Journal of Atmospheric Sciences, 25:* 889–899.

Mrose, H., 1966. Measurements of pH and chemical analysis of rain, snow and fog water, *Tellus, 28:* 266.

Munn, R. E., and Bolin, B., 1971. Global air pollution-meteorological aspects, *Atmospheric Environment, 5,* 363.

National Air Pollution Control Administration (NAPCA), 1970. Nationwide inventory of air pollutant emissions, 1968 (Raleigh, North Carolina: NAPCA), Publication No. AP-73.

Oddie, B. C. V., 1962. The chemical composition at cloud levels, *Quarterly Journal of the Royal Meteorological Society, 88:* 535.

Pales, J. C., and Keeling, C. D., 1965. The concentration of atmospheric carbon dioxide in Hawaii, *Journal of Geophysical Research, 70.*

Parkin, D. W., Phillips, D. R., Sullivan, R. A. L., and Johnson, L., 1970. Airborne dust collections over the North Atlantic, *Journal of Geophysical Research, 75:* 1782–1793.

Peterson, J. T., and Junge, C. G., 1971. Sources of particulate matter in the atmosphere, *Man's Impact on the Climate,* edited by W. H. Matthews, W. W. Kellogg, and G. D. Robinson (Cambridge, Massachusetts: The M.I.T. Press), 310–320.

Plass, G. N., 1956. The carbon dioxide theory of climatic change, *Tellus, 8:* 140–156.

Quenzel, H., 1970. Determination of size distribution of atmospheric aerosol particles from spectral solar radiation measurements, *Journal of Geophysical Research, 75:* 2915–2921.

Rasool, S. I., and Schneider, S. H., 1971. Atmospheric carbon dioxide and aerosols: effects of large increases on global climate, *Science, 173:* 138–141.

Reiquam, H., 1970. European interest in acidic precipitation, *Precipitation Scavenging,* Proceedings of symposium held at Richland, Washington, June 2–4, 1970, sponsored by Pacific Northwest Laboratory, Battelle Memorial Institute, and the Fallout Studies Branch, Division of Biology and Medicine, U.S. Atomic Energy Commission, AEC Symposium Series No. 22 (Washington, D.C.: U.S. Atomic Energy Commission), *261,* p. 289.

Reiter, E. R., 1961. *Meteorologie der Strahlstürme* (Vienna: Springer-Verlag).

Robinson, G. D., 1958. Some observations from aircraft of surface albedo and the albedo and absorption of cloud, *Archiv für Meteorologie, Geophysik in Bioklimatologie, Seve B, 9:* 28–41.

Robinson, G. D., 1971. Particles in the atmosphere and the global heat budget (unpublished).

Robinson, E., and Robbins, R. E., 1971. Emissions, concentrations, and rate of particulate atmospheric pollutants, Final Report, SRI Project, SCC-8507.

Shannon, L. J., Vandegrift, A. E., and Gorman, P. G., 1970. Assessment of small particle emission (less than 2 micron) (Kansas City, Missouri: Midwest Research Institute), Report 425.

Soulage, G., 1958. Contribution des fumées industrielles à l'enrichessement de l'atmosphère en noyaux glacogènes, *Bulletin de l'Observatoire du Puy de Dôme, 4:* 121–124.

Squires, P., and Twomey, S., 1960. The relation between cloud droplet spectra and the spectrum of cloud nuclei, Physics of Precipitation, American Geophysical Union, *Geophysics Monographs,* No. 5: 211–219.

Study of Critical Environmental Problems (SCEP), 1970. *Man's Impact on the Global Environment* (Cambridge, Massachusetts: The M.I.T. Press).

Telegadas, K., 1971. The seasonal atmospheric distribution and inventories of excess carbon[14] from March 1955 to July 1969 (Washington, D.C.: U.S. Atomic Energy Commission), Report HASL-243.

Telford, J. W., 1960. Freezing nuclei from industrial processes, *Journal of Meteorology, 17:* 676–679.

Twomey, S., Jacobowitz, H., and Howell, H. B., 1968. Comments on anomalous cloud lines, *Journal of Atmospheric Sciences, 25:* 333–334.

Twomey, S., and Warner, J., 1967. Comparison of measurements of cloud droplets and cloud nuclei, *Journal of Atmospheric Sciences, 26:* 702–703.

Twomey, S., and Wojciechowski, T. A., 1969. Observations of the geographical variation of cloud nuclei, *Journal of Atmospheric Sciences, 26:* 684–688.

Van Valin, C. C., 1971. Effects on the stratosphere of SST operation (Washington, D.C.: U.S. Government Printing Office), NOAA Technical Report ERL 208–APCL 21, p. 13.

Vittori, O., and Prodi, V., 1970. Particle-free space in Stefan flow, *Precipitation Scavenging,* Proceedings of symposium held at Richland, Washington, June 2–4, 1970, sponsored by Pacific Northwest Laboratory, Battelle Memorial Institute, and the Fallout Studies Branch, Division of Biology and Medicine, U.S. Atomic Energy Commission, AEC Symposium Series No. 22 (Washington, D.C.: U.S. Atomic Energy Commission), p. 403.

Warner, J., and Twomey, S., 1967. The production of cloud nuclei by cane fires and their effect on cloud droplet concentration, *Journal of Atmospheric Sciences, 25:* 704–706.

Went, F. W., 1960. Organic matter in the atmosphere, *Proceedings of the National Academy of Sciences, 46:* 212.

Whitby, K. T., and Clark, W. G., 1967. Electric aerosol particle counting and size distribution measuring system for the 0.015 to 1μ size range, *Tellus, 18:* 573–586.

Woodcock, A. H., 1953. Salt nuclei in marine air as a function of altitude and wind force, *Journal of Meteorology, 10:* 362–371.

Yamamoto, G., and Tanaka, M., 1969. Determination of aerosol size distribution from spectral attenuation measurements, *Applied Optics, 8:* 447–453.

Yamamoto, G., and Tanaka, M., 1971. Increase of global albedo due to air pollution (to be published).

Yamamoto, G., Tanaka, M., and Kamitani, K., 1966. Radiative transfer in water clouds in the 10 micron window region, *Journal of Atmospheric Sciences, 23:* 305–313.

Yamamoto, G., Tanaka, M., and Arao, K., 1971. Secular variation of turbidity coefficient over Japan (unpublished).

9 Modification of the Stratosphere

9.1 Introduction

9.1.1 Importance of the Stratosphere

Three properties of the stratosphere make natural and artificial pollution of this layer more critical than that of the troposphere. First, the thermal stability and virtual cloudlessness of the stratosphere permit many pollutants to reside longer in the stratosphere (for a time measured in years) than in the troposphere (for a time measured in days or weeks). The equilibrium concentration produced by a continuous source can thus be ten to one hundred times greater in the stratosphere than in the troposphere for an equal source. Second, ultraviolet radiation creates ozone that is important both in controlling stratospheric temperatures and in shielding life at the earth's surface from undesirable ultraviolet radiation. It will be seen (Section 9.4) that even small changes of pollutants, measured in parts per million (ppm), like water vapor to parts per billion (ppb, 10^{-9}), like nitric oxide can play a role in the ozone equilibrium. Finally, the lower stratosphere is close to radiative balance. That is, the individual heating and cooling rates due to water vapor, carbon dioxide, and ozone very nearly cancel each other.

9.1.2 Observed Fluctuations

Measurements of water vapor indicate fluctuations with periodicities or trends of more than a few years. For example, Figure 9.1 shows the observed mixing ratios of stratospheric water vapor over Washington, D.C. It contains a variability from month to month, but between 1964 and 1969 an upward trend appears to be present.

Total ozone observations at a number of stations over the globe also show fluctuations of various periods. The trend of total ozone during the 10-year period for some of these stations is shown on Figure 9.2. Some stations (not shown here) exhibited a down-

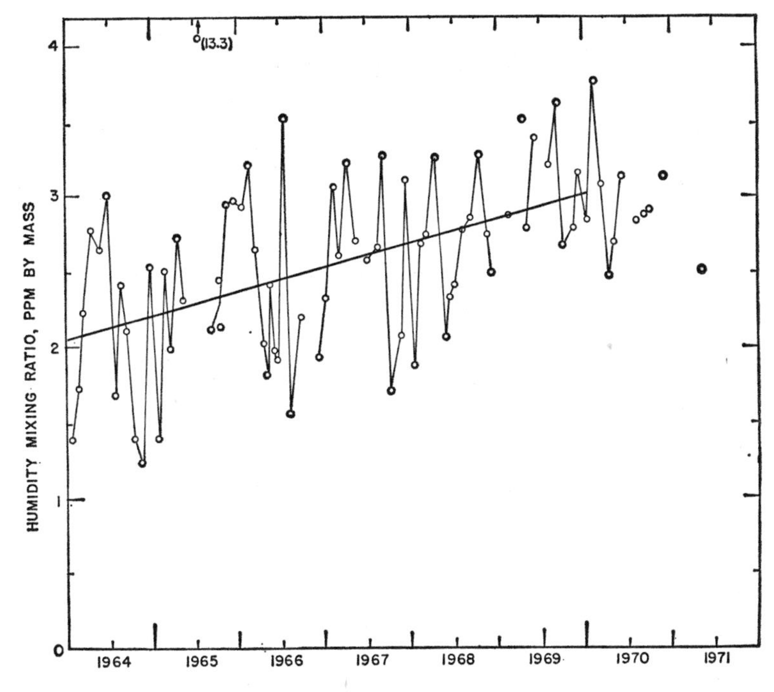

Figure 9.1 Water vapor mixing ratios in the stratosphere at 50 millibars or 20.6 kilometers near Washington, D.C., United States. The straight line is a best-fit line between 1964 and 1969. Circles in adjacent months have been connected by straight lines. The upward trend between 1964 and 1969 appears to have leveled off after early 1970.
Source: Mastenbrook, 1971.

ward trend during the same period, and in many cases (at Arosa, for example) these trends appear as part of a longer period fluctuation. Nevertheless, the general increase in total ozone for the stations shown in Figure 9.2 seems to be representative of a global pattern.

Insofar as can be determined, the upward trend in water vapor between 1964 and 1969 and the upward trends in ozone in the 1961 to 1970 decade are due to natural causes. The natural fluctuations in stratospheric constituents will make it difficult to assign man-made origins to future trends. But equally disturbing is our inability to explain the observed changes now in progress.

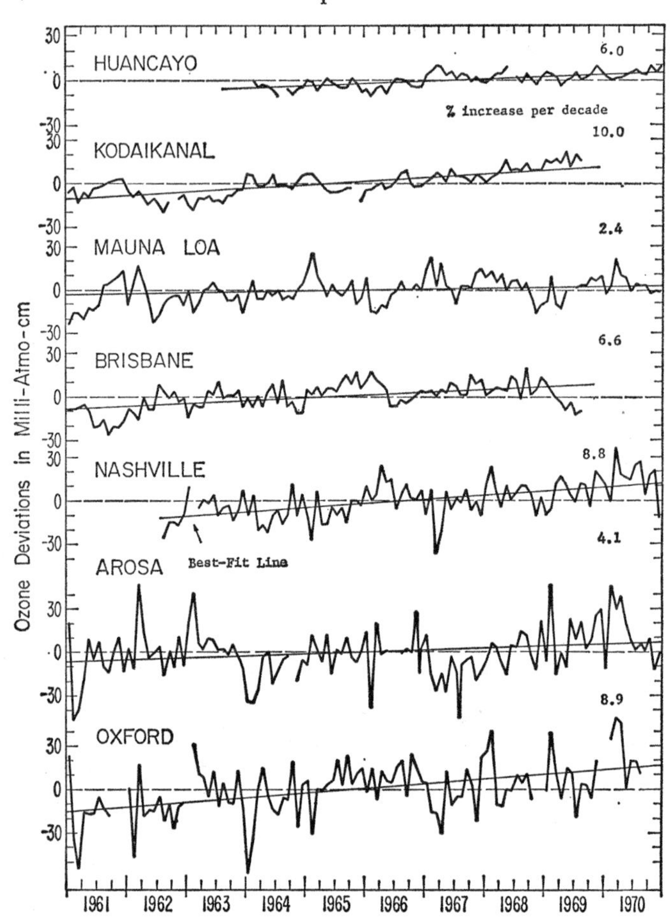

Figure 9.2 Total ozone trends for selected world stations expressed as departures from normal between 1958 and 1970. Other stations show an increase similar to that in this figure while some exhibit no apparent trends or decreasing trends. Before 1960, when fewer stations were in operation, both increasing and decreasing total ozone trends of about the same magnitude as in the figure were observed.
Source: Komhyr et al., 1971.

It is crucial that measurements of all possible climate-related constituents be made to obtain both short- and long-term fluctuations.

9.1.3
Tropospheric Transport to the Stratosphere

Exchange of air between troposphere and stratosphere can and does alter the constituents of these two layers of the atmosphere. Transport of ozone and radioactive constituents from stratosphere to troposphere has received considerable attention, but neither is likely to involve significant climatic effects when in the troposphere. The transfer of many constituents from troposphere to stratosphere deserves consideration for many reasons to be discussed in later sections.

Four mechanisms have been proposed whereby tropospheric air transfers into the stratosphere:

1. Upward movements usually thought to occur in equatorial latitudes.
2. Quasi-horizontal circulation through the tropopause level in the vicinity of the tropopause gap.
3. Upward mixing at all latitudes.
4. Convective updrafts associated with intense thunderstorms whose clouds penetrate into the stratosphere.

The dryness of the stratosphere led Brewer (1949) to argue that the first of these mechanisms was most important, since stratospheric frost point readings of $-80°$ to $-85°C$ suggest that air has been cooled to this temperature, which is the temperature of the equatorial tropopause. Mastenbrook (1971) finds a seasonal variation in stratospheric moisture up to about 20 kilometers over Washington, D.C., with a summertime rise that suggests a thunderstorm source for these altitudes in the summer season. The thunderstorm mechanisms wherever they occur involve upward transport of tropospheric constituents within ice crystal clouds.

9.1.4
Layers of the Upper Atmosphere

The atmosphere above the tropopause contains several layers of which the first, the stratosphere, is of greatest present concern (see Figure 9.3).

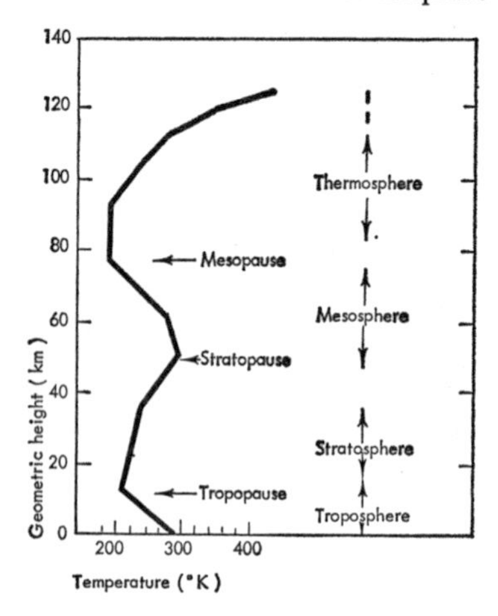

Figure 9.3 Nomenclature for regions of the upper atmosphere, based on temperature distributions.

9.2
Stratospheric Injections by High-Flying Aircraft

The SCEP Report (1970) treated the climatic consequences of injecting pollutants into the stratosphere by a fleet of commercial transport aircraft. The assumptions on which that analysis was based in July 1970 are no longer valid. No estimates are presently available for the fleet size and operations or of engine emissions for either the Concorde or the Tupolev-144. Although quantitative estimates of the future stratospheric concentrations of pollutants cannot be made, in this section the SCEP climatic analysis will be reexamined.

A number of assumptions were required in the SCEP Report to provide quantitative estimates of stratospheric contamination by a fleet of supersonic aircraft. The fleet was then estimated at 500 planes of which 334 were large 4-engine United States aircraft and 166 smaller aircraft with engines emitting the equivalent of 2 engines of U.S. type. An engine consumes 33,000 lb of fuel per hour. Each aircraft would fly an average of 7 hours per day at 20 kilometers (65,000 feet) at a speed of Mach 2.7. The ratio of exhaust emission to fuel consumption is

CO_2:3 (that is, 3 pounds of CO_2 to 1 pound of fuel consumed)
H_2O:1.3
CO:0.04
NO:0.04
SO_2:0.001
Soot:0.0001

Since the SCEP Report, some modifications to the emission rates have been reported by the U.S. engine manufacturer. For example, the newer estimates for the NO emissions are several-fold smaller than in the SCEP Report. Later estimates of the fleet size, on the other hand, have suggested as many as 1000 aircraft by the same period 1985–1990.

Each of the major interpretations of the SCEP Report dealing with potential effects of a fleet of supersonic aircraft is reviewed now in the light of present knowledge.

1. The radiative effects of carbon dioxide, water vapor, and the minor exhaust products are not likely to be significant in the light of the much larger sources of combustion products in the troposphere.
2. The SCEP analysis, assuming fuels with current average sulfur contents, indicated that the particle concentrations in the stratosphere would equal or locally exceed the natural background. In the region of peak flight activity the concentrations would be of the order of the particle concentration found in the stratosphere in 1968 after the Mt. Agung volcanic eruption. The largest contribution to the particle mass comes from the sulfur in the fuel. It is feasible, however, to use fuels with much smaller sulfur contents than those currently used. A detailed analysis of the Mt. Agung volcano effects in the stratosphere is discussed in Section 9.6. From these recent data no new conclusions have been drawn that would alter the SCEP analysis. If the addition of particles from high-flying aircraft increases the background value by more than tenfold, it would probably warm the stratosphere by a detectable amount. However, whether or not a significant man-made particle increase would occur depends on the fleet size and operation, engine emission, and the sulfur content of the fuel.
3. Water vapor is a substance that may increase or decrease in the stratosphere due to either man's or nature's activities. Changes

in stratospheric moisture may then alter the frequency, thickness, or distribution of clouds above the tropopause (all of which are rarely observed phenomena).

The stratosphere is generally very dry, 2 to 3 ppm by mass (Mastenbrook, 1971). Relatively large amounts of moisture are necessary to saturate it, but there are two regions of the lower stratosphere in which the air is relatively close to saturation because the temperatures may be colder than −75° or −80°C. These are the regions just above the equatorial tropopause and parts of the polar region in winter at altitudes generally between 25 and 35 kilometers. There has been only one case in which persistent contrails were reported behind aircraft flying above 15 kilometers. This was over the equatorial tropopause (Los Alamos National Laboratory, unpublished 1970).

Some natural clouds, sometimes called mother-of-pearl clouds, have been reported in the lower stratosphere of the polar night over both Antarctica and Norway. In the Northern Hemisphere Norway lies on the eastern edge of the cold region; reports from that country average about one or two sightings each of the three winter months. It is not clear whether more such stratospheric clouds would be present in the region north of Iceland where the temperature is lower than over Norway or whether the Norwegian sightings reflect the orographic lifting over the mountains of that country. For reasons not now understood, no mother-of-pearl sightings were reported in Norway between 1892 and 1925 (Hesstvedt, 1958).

It is presumed that there will be a direct relationship between changes in stratospheric moisture and stratospheric cloudiness in the two sensitive parts of the atmosphere, but there has been no advance in knowledge which allows change in the conclusions of the SCEP Report concerning stratospheric cloudiness. It is still uncertain whether stratospheric cloudiness would increase or what the consequences to the climate of such an increase would be.

9.3
Stratospheric Transmission of Ultraviolet Radiation

Aircraft engines emit water vapor and nitric oxide, both of which could conceivably produce changes in the amount of stratospheric ozone. Since ozone is also one of the radiatively active gases in the atmosphere, particularly in the stratosphere, climatic modifications may result. It still appears likely, in support of the SCEP judgment, that the radiative effects due to possible changes in ozone would be negligible. There is another possible consequence of the potential change by man of the stratospheric ozone content, one that is due to the ultraviolet shielding properties of ozone. Variations in ultraviolet radiation reaching the earth's surface could have serious consequences for the biosphere in general and man in particular.

The intensity of incoming solar ultraviolet radiation received at the ground depends on the intensity of the radiation at the top of the atmosphere, depleted primarily by absorption by ozone but also by scattering by air molecules and dust. Although the undepleted solar radiation contains decreasing energy with decreasing wavelength below 3400 Å, the large increase of absorptivity by ozone below 3400 Å results in an even more pronounced decrease of the solar radiation received at the ground, so that very little of the solar radiation below approximately 3100 Å penetrates through the atmosphere to the surface (London, 1969).

It is well known that the human skin is very sensitive to various wavelengths of ultraviolet light; the peak of this sensitivity is at a wavelength of about 2950 Å. Multiplication of relative sensitivity values by the observed distribution of the solar radiation received at the ground results in an actual "action spectrum" of the biologically effective solar ultraviolet radiation. Such a curve is shown in Figure 9.4 for Davos, Switzerland (after Urbach, 1969). The "action spectrum" curve for erythema rises sharply from about 2950 Å to a peak at 3075 Å. Although the peak is quite narrow, the curve shows some extension to 3300 Å. Higher total ozone values, as might be found at more northerly latitudes, would shift the peak in the action spectrum slightly to longer wavelengths; lower total ozone would shift the peak toward shorter wavelengths. Qualitatively, however, the shape of the curve would remain about the same. The maximum in the actual action spectrum curve occurs in that spectral region where very little ultraviolet radiation penetrates to the surface but where the amount that does penetrate depends sensitively on the total ozone present

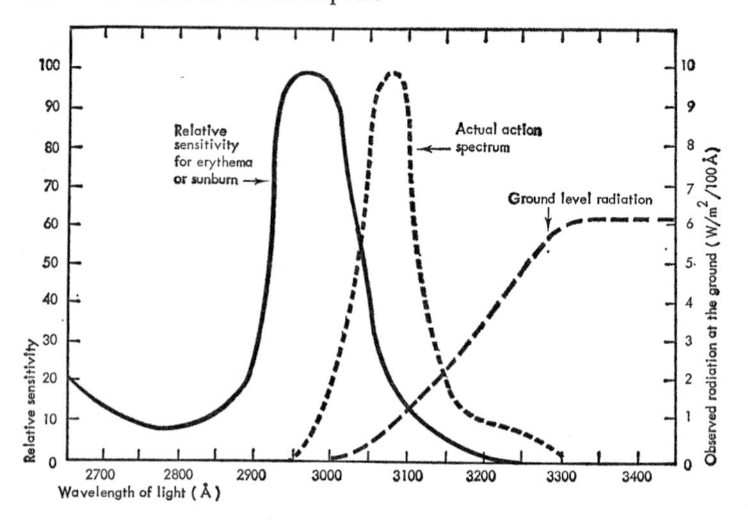

Figure 9.4 Actual action spectrum of biological efficiency of solar ultraviolet radiation for Davos, Switzerland. The actual action spectrum is the product at each wavelength of the relative sensitivity and the ground level radiation. Source: After Urbach, 1969.

in the path of the sun's rays. Large amounts of ozone and/or long atmospheric paths (low sun) would reduce the ultraviolet radiation transmitted to the ground. Small amounts of ozone and/or short paths (high sun) increase the ultraviolet transmitted to the ground.

It has been suggested (see for instance, Blum, 1959; Urbach, 1969) that the apparent latitudinal increase in the incidence of skin cancer toward the equator (for example, Auerbach, 1961) is because of the increase of ultraviolet radiation received at the ground resulting from the higher average solar elevation angle and decreased observed total ozone toward the equator; for a fixed wavelength and solar zenith angle a change in total ozone will result in a relative change in ultraviolet radiation received at the surface. Calculated results for such relative changes as a function of wavelength have been made for an assumed solar zenith angle of 30° and an assumed 1 percent change in total ozone. These are shown in Table 9.1 for three different ozone amounts representing small, average, or large (corresponding to equatorial, mid-latitude, or polar average ozone values, respectively).

Table 9.1
The Percentage Change of Ultraviolet Radiation Received at the Ground for a 1 Percent Change in Three Assumed Values of Total Ozone and an Elevation Angle (z) of 30°

Wavelength of Ultraviolet Radiation (Å)	Three Assumed Values of Total Ozone (cm) 0.250	0.350	0.450
	Percentage Change of Radiation Received at the Ground (percent)		
2950	6.0	8.2	11.5
3000	3.0	4.0	5.2
3050	1.5	2.0	2.7
3100	0.7	1.0	1.4
3150	0.4	0.6	0.8
3200	0.1	0.25	0.4
3250	0.05	0.1	0.2
3300	0.01	0.05	0.1

Note: Computations were made using an assumed mixed absorbing-scattering atmosphere. The unit of concentration is centimeters, representing the equivalent depth of a volume of unity cross section if all the ozone molecules were brought to standard temperature and pressure.
Source: John De Luisi, National Center for Atmospheric Research, Boulder, Colorado, United States (private communication).

It can be seen from Table 9.1 that the relative increase in ultraviolet radiation increases with decreasing wavelength and with increasing assumed total ozone amount. In both cases, of course, the absolute value of the incident radiation received at the ground in the interval 3050 Å to 3100 Å would be most significant. The maximum ozone concentration is found in the lower stratosphere, and injections by high-flying aircraft could conceivably modify the ozone balance at these levels. The sensitivity of fundamental life processes to variations in ultraviolet radiation (see, for example, Trosko, Krause, and Isoon, 1970) requires careful study of the distribution of atmospheric ozone and of possible changes in the ozone distribution resulting from man's activities such as aircraft flights in the stratosphere.

9.4 Stratospheric Ozone in an Oxygen-Nitrogen-Hydrogen Atmosphere

9.4.1 The Oxygen-Ozone System

Ozone was identified as a nontrivial constituent of the atmosphere about 100 years ago, and its presence was first correctly explained by Sydney Chapman (1930, 1943). He pointed out that ozone is present in relatively large quantities in the stratosphere primarily as the result of photodissociation of molecular oxygen and subsequent recombination of atomic and molecular oxygen to form ozone. Ozone is, in turn, destroyed either by photodissociation or by recombination with atomic oxygen. The complete scheme proposed by Chapman for the equilibrium ozone distribution consists of a set of five photochemical reactions involving atomic oxygen, O, molecular oxygen, O_2, and ozone, O_3. This scheme reproduces the general shape and order of magnitude of the observed ozone distribution (see, for instance, Craig, 1965). There are, however, important differences between observed distributions and those derived from the Chapman "pure oxygen" scheme. The present discussion is limited to stratospheric conditions (15 to 50 km).

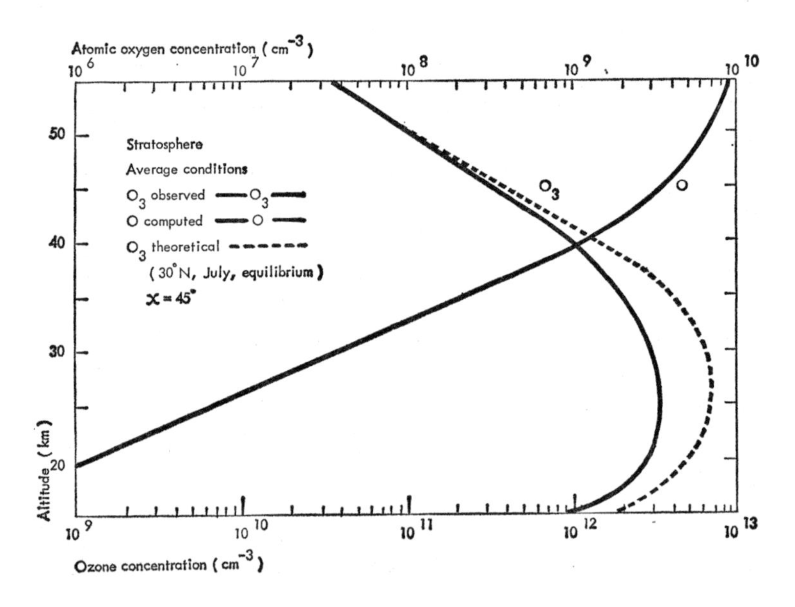

Figure 9.5 Average values of the observed concentration of ozone and the associated concentration of atomic oxygen in a sunlit atmosphere compared with equilibrium conditions for a "pure oxygen" atmosphere for July at 30° N. Solar zenith angle $\chi = 45°$.
Sources: London and Park, 1971, unpublished, and Nicolet, 1970a.

Figure 9.5 compares the theoretical photochemical vertical ozone distribution for a "pure oxygen" atmosphere with an average observed distribution. In the lower stratosphere the observed curve departs significantly from the photochemical equilibrium distribution. The difference is due primarily to the length of time that is involved in the production of oxygen atoms (and, therefore, ozone formation). The production time increases markedly with decreasing altitude below 50 kilometers and with increasing solar zenith angle. Thus, for instance, the production time (that is, the time required to produce an ozone concentration equivalent to the average observed value shown in Figure 9.6) increases for a zenith angle of 60° from about 2 days at 40 kilometers to 10 days at 30 kilometers and about 1 year at 20 kilometers. This is shown in Figure 9.6, where the restoration time to the observed values for two different zenith angles are plotted as functions of height. In the upper stratosphere (35 to 50 km) there is approximate photochemical equilibrium, since the production times involved are less than one or two days. In the lower stratosphere

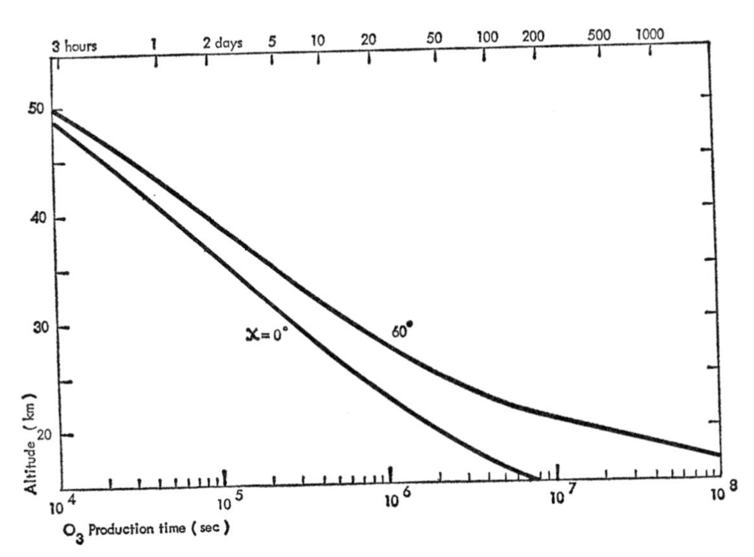

Figure 9.6 The ozone production time (to the observed value) by photodissociation of molecular oxygen as a function of height in the stratosphere for a zenith angle $\chi=0°$ and 60°.
Source: Derived from data given in Nicolet, 1970a.

(below 25 km) with much longer restoration times, atmospheric transport plays the dominant role in determining the vertical and horizontal ozone distribution. The atomic oxygen vertical distribution is determined by calculation from the stratospheric condition of photochemical equilibrium between O and O_3, which always exists in a sunlit atmosphere. The atomic oxygen concentration is always less than that of ozone in the stratosphere and becomes negligible after sunset. The calculated values for atomic oxygen, corresponding to the observed ozone distribution shown in Figure 9.6, increase from 10^6 cm^{-3} * at 20 kilometers to 6.5×10^9 cm^{-3} at 50 kilometers.

Any theoretical model for a pure oxygen atmosphere based on photochemical equilibrium conditions in the stratosphere shows, in general, that the calculated ozone concentration is larger than the observed value. At high latitudes, however, in the upper troposphere and lower stratosphere, the reverse is generally true (London, 1968). Thus, a steady-state distribution of ozone and atomic oxygen as functions of altitude and latitude in the stratosphere must involve other factors, such as atmospheric transport, to explain the observed ozone distribution in the major part of the stratosphere. Some preliminary attempts have been made to study the effects of atmospheric motions on the ozone distribution in a "pure oxygen" photochemical model (see, for instance, Prabhakara, 1963).

9.4.2 Nitrogen Oxides

Nitrogen oxides participate in the chemical processes for stratospheric ozone, as pointed out by Nicolet (1965). NO_x (that is, NO and/or NO_2) have their source regions either in the mesosphere (as NO) or in the lower troposphere (chiefly as NO_2). Nicolet (1970b) has suggested, following the study of Greenberg and Heicklen (1970), that excited atomic oxygen produced by photodissociation of ozone could also interact with nitrous oxide (N_2O) to produce nitric oxide locally in the stratosphere. The concentration of NO in the stratosphere is not well known, although Ackerman and Frimout (1969) from their infrared spectroscopic observa-

* Stands for 10^6 particles cm^{-3}, where particles may be atoms or molecules.

tions suggest an upper limit of the NO_2 mixing ratio in the lower stratosphere of 3×10^{-8}.

Nitrogen dioxide can be photodissociated by near-ultraviolet sunlight to give atomic oxygen and nitric oxide, or it can combine with atomic oxygen to produce nitric oxide and molecular oxygen. The nitric oxide, in turn, combines with ozone and, in the ozone destruction process, forms molecular oxygen and reforms nitrogen dioxide. Thus, in the stratosphere the photochemical effect of NO leads to a destruction of ozone through catalytic action. The photochemical equilibrium of the NO to NO_2 ratio is always close to unity in the lower stratosphere but increases rapidly above about 30 kilometers.

As in the case of a "pure oxygen" atmosphere, it is necessary to consider the ozone loss in the presence of NO (and OH, which will be discussed in Section 9.4.3). This is shown in Figure 9.7,

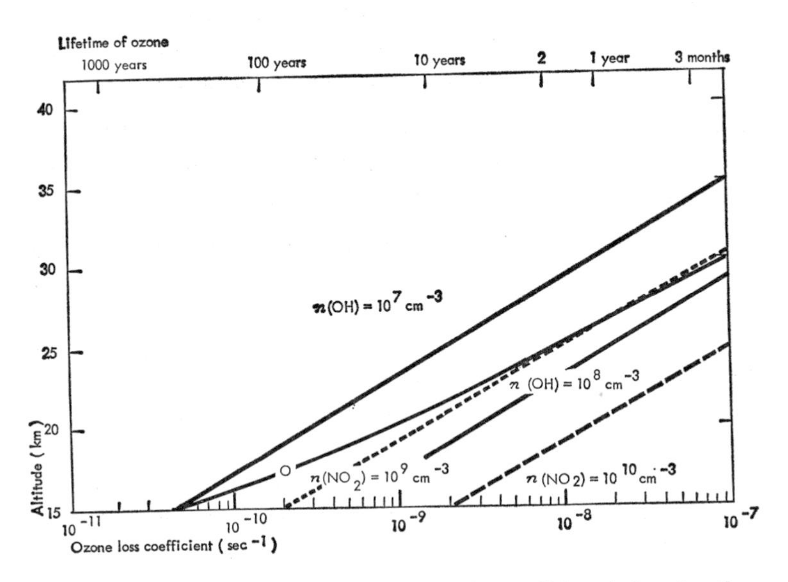

Figure 9.7 The height variation of the ozone loss coefficient (relaxation time to reduce the observed ozone concentration to $1/e$ of its initial value) as a function of the NO_2 density and OH density as compared with the loss due to recombination with atomic oxygen.
Source: Derived from data given in Nicolet, 1970a.

where the ozone loss coefficient is plotted as a function of altitude for two different values of the NO_2 density. The ozone loss coefficient is the inverse of the relaxation time, or the time required for perturbed ozone to be reduced to $1/e$ *of* its initial value. The normal loss process for ozone in a "Chapman" atmosphere is recombination with atomic oxygen; that is, $O + O_3 \rightarrow 2\ O_2$. Therefore, a curve of the ozone loss coefficient for this recombination process with atomic oxygen is given for comparison with the other curves in Figure 9.7. It is seen that NO_2 concentrations of the order of 10^{10} cm^{-3} (5 parts per billion by volume, ppbv, 10^{-9}) at 20 kilometers produce a 1-year relaxation time, and that for an NO_2 concentration of 10^9 cm^{-3} (0.5 ppbv, 10^{-9}) at that height the relaxation time of course is longer (of the order of 10 years). For a nitrogen dioxide concentration of 10^9 cm^{-3}, the loss coefficient is the same as that for atomic oxygen in the region 20 to 30 kilometers. In other words, the lower stratosphere is also characterized by relatively slow action of NO on ozone. For example, if we assume an unperturbed mixing ratio of 3×10^{-9}, or 3 ppbv, 10^{-9} in the stratosphere, the relaxation times for ozone are

Height (km)	15	20	25	30
NO (cm^{-3})	1.2×10^{10}	5.1×10^9	2.2×10^9	1.1×10^9
τ (years)	13	5	0.8	0.55

The long lifetimes at 20 kilometers and below show that equilibrium conditions do not adequately describe the NO-ozone photochemistry in the lower stratosphere.

Johnston (1971) has recently computed the effect of NO (both natural and that due to exhaust products of stratospheric aircraft) on the ozone distribution in the stratosphere. His results are based on photochemical equilibrium conditions using a simple model that attributes all differences between observed and theoretical (pure oxygen) distributions to the effects of atmospheric NO. But as pointed out by Crutzen (1970, 1971), who has considered the oxygen-hydrogen-nitrogen reactions in some detail, the problem of the stratospheric ozone distribution is very complex and must be studied in an atmospheric model in which proper treatment is given to dynamical processes before any definitive results can be considered reliable.

The effectiveness of the NO catalytic reactions is reduced by the slow destruction of NO and NO_2 in recombination with OH and HO_2 (hydroperoxyl) to form nitrous or nitric acid in the stratosphere, but the absolute values of the rate coefficients of these reactions are not well known. Nitric acid has been observed in the stratosphere (Murcray et al., 1969) and analysis of these data (Rhine, Tubbs, and Williams, 1969) suggests a value at 20 km of 5 ppbv (10^{-9}).

9.4.3 Hydrogen Compounds

The basic theory describing how water vapor plays a role in the photochemistry of ozone is well known (Bates and Nicolet, 1950). One of the products of photodissociation of ozone at wavelengths below 3100 Å is excited atomic oxygen. Excited atoms react very rapidly with water vapor to produce two hydroxyl molecules (OH). Of less importance in the stratosphere is the photodissociation of water vapor that also produces OH. Hydrogen, hydroxyl, and hydroperoxyl radicals and hydrogen peroxide molecules are linked in the catalytic destruction of atomic oxygen and ozone. Part of this reaction scheme is shown diagrammatically in Figure 9.8. The catalytic reactions of $OH + O_3 \rightarrow HO_2 + O_2$ and $HO_2 + O_3 \rightarrow OH + 2\ O_2$ have been suggested by many as representing a major destructive mechanism for ozone in the lower stratosphere (see, for instance, Hampson, 1966; Hunt, 1966; Dütsch, 1968; Leovy, 1969; Hesstvedt, 1968; Crutzen, 1969). However, recent experimental analysis (Langley and McGrath, 1971), following the earlier studies of Kaufman (1964) and DeMore (1967), tend to support the views expressed by Nicolet (1970) that such reactions

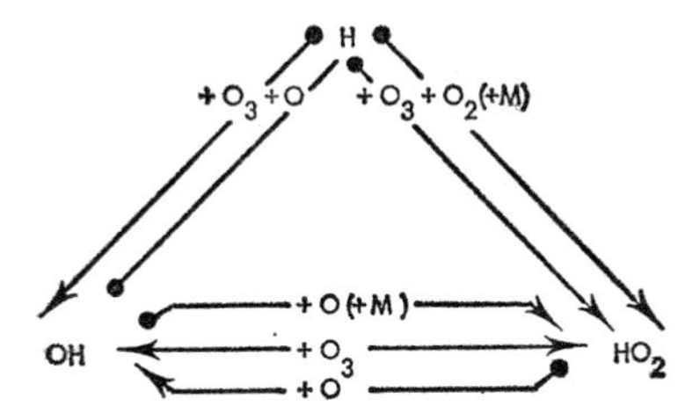

Figure 9.8 Schematic diagram showing the photochemical interaction of H, OH, and HO_2 in the stratosphere.

seem to be very slow at pressures and temperatures of the lower stratosphere and probably do not play an important role in the catalysis of ozone at these levels. Thus, the principal effect of the hydrogen compounds in the stratosphere is probably through the reactions of OH and HO_2 with atomic oxygen that lead to the reformation of molecular oxygen. As indicated in Figure 9.7 a concentration of the OH radical of the order of 10^8 cm^{-3} or less at 20 kilometers would result in an ozone relaxation time of more than 3 years. Changes in water vapor concentration, even by a factor of 2 from the observed values, would probably have little direct effect on the ozone content of the lower stratosphere. Since a major part of the stratospheric ozone lies in this region (20 to 30 km), the change in total ozone will in all likelihood also be small.

Several attempts have been made to include the oxygen-water vapor photochemical system in a dynamic model of the stratosphere (Hunt, 1969; Gebhart, Bojkov, and London, 1970) but the dynamical considerations were highly simplified and the photochemical schemes need modification in light of the suggested slow reaction rates as discussed earlier.

9.4.4 Conclusions

Stratospheric ozone is formed as a result of photodissociation of molecular oxygen. It is destroyed either by photodissociation or by recombination with atomic oxygen. Nitrogen oxides and hydrogen compounds may play a role in the stratospheric chemistry leading to the removal of ozone. Their photochemical influence in reducing ozone in the upper stratosphere is relatively direct. In the lower stratosphere, however, atmospheric transport, both vertical and horizontal, as well as the interaction between nitrogen oxides and hydrogen compounds, represents controlling factors in variations of the ozone distribution.

It is likely that significant modification by man of the existing "natural" composition of the lower stratosphere would be required to produce a decrease in stratospheric ozone. Because of the complexities of the problem, involving the interactions of stratospheric photochemistry and atmospheric dynamics, it is not possible to say, at present, how large this decrease would be or how this would affect the total ozone in a vertical column. Because of its great potential importance, considerable research effort should be undertaken to minimize this uncertainty before man significantly contaminates the stratosphere.

9.5 Sources and Sinks of Trace Gases Other than Ozone

9.5.1 Introduction

Measurements of gases in the stratosphere are still rather rudimentary and often a matter of uncertainty and controversy. Trace gases may, in part, control the concentrations of two climatically important constituents of the stratosphere: ozone and water vapor. Of course, the two principal constituents, molecular nitrogen and oxygen, and the noble gases, helium, argon, and neon, maintain virtually constant mixing ratios in both the troposphere and stratosphere. The same constant fractional concentration is generally assumed for carbon dioxide (CO_2) up to thermospheric levels.

9.5.2 Carbon Dioxide

Because of the relatively long atmospheric residence time of a CO_2 molecule before it enters the biosphere or ocean, the exchange between the troposphere and lower stratosphere can produce almost equal concentrations in the two layers (Bischof and Bolin, 1966; Georgii and Jost, 1969; Jost, 1971). However, it would be valuable to augment the present stratospheric observations obtained aboard commercial aircraft by additional data at higher altitudes.

On the basis of knowledge of the oxygen absorption and of the CO_2 absorption cross section, it is known that no direct dissociation can occur in either the stratosphere or mesosphere. The dissociation coefficients range from 10^{-8} sec^{-1} at the mesopause level to 10^{-14} sec^{-1} at the tropopause. On the other hand, CO_2 will dissociate in the thermosphere to produce carbon monoxide (CO) and atomic oxygen (O). Another possibility that may play a small role in destroying CO_2 in the stratosphere is the reaction with CO_2 of O atoms in their first excited state (Preston and Cvetanovic, 1966). This reaction is still uncertain.

9.5.3
Carbon Monoxide

There are three possible sources of carbon monoxide, CO, for the stratosphere; the injection of CO from the troposphere by upward motions (Junge, Seiler, and Warneck, 1971), downward transport from the photodissociation of CO_2 in the thermosphere (Hays and Olivero, 1970), and the direct formation within the stratosphere. Seiler and Junge (1969) observed a sharp decrease of CO concentration moving upward through the tropopause. This decrease is certainly partly due to a sink of CO in the stratosphere (Pressman and Warneck, 1970; Hessvedt, 1970) by the reaction of carbon monoxide with a hydroxyl radical, OH, leading to the formation of CO_2. The product of the rate coefficient of the order of 10^{-3} cm^3 sec^{-1} (Greiner, 1970) with OH concentrations from $5 \times 10^{+6}$ to $5 \times 10^{+7}$ cm^{-3} (Nicolet, 1970) yields the relatively large loss coefficients for CO of 5×10^{-7} and 5×10^{-6} sec^{-1}. In view of the source of CO due to photodissociation of CO_2 above the stratosphere it would be desirable to measure CO in the upper stratosphere.

9.5.4
Water Vapor

Water vapor, H_2O, provides a source of OH and hydroperoxyl radicals, HO_2, which are of great importance in the behavior of stratospheric and mesospheric ozone (Bates and Nicolet, 1950). The reactions involving these constituents require a knowledge of the latitudinal and vertical distribution of water vapor. Present estimates of the average stratospheric mixing ratio lie in the range 2.5 to 3.0×10^{-6} (Williamson and Houghton, 1965; Mastenbrook, 1971; Sissenwine, Grantham, and Samela, 1968; McKinnon and Morewood, 1970). Stratospheric fluctuations of moisture are dependent mainly on upward transport of moisture from the troposphere where the concentrations are many orders of magnitude greater than those in the stratosphere. To a far smaller extent, it is also related to the stratospheric content of methane, CH_4, and molecular hydrogen, H_2.

9.5.5
Methane

The behavior of methane is relatively simple in the stratosphere since its sole source is transport from the troposphere and no re-formation is possible when it dissociates. A continuous daytime reaction between excited oxygen atoms (Cadle, 1964) and CH_4 destroys CH_4 in the stratosphere, with a loss coefficient as high as 10^{-7} sec^{-1} in the upper stratosphere (Nicolet, 1970). In the mesosphere, the photodissociation by the solar line Lyman-α is the principal destruction process.

The mixing ratio of CH_4 in the troposphere is 1.5×10^{-6}. Measurements of CH_4 (Bainbridge and Heidt, 1966; Scholz et al., 1970) show a decrease of mixing ratio from the lower stratosphere to the stratopause. The loss of CH_4 in the stratosphere leads ultimately to the formation of one CO_2 and two H_2O molecules. This addition of water vapor to the content in the stratosphere (Bates and Nicolet, 1950b) depends on the upward transport of CH_4 through the tropopause level.

9.5.6
Molecular Hydrogen

The behavior of H_2 is more complicated than that of CH_4. The oxidation of H_2 by excited oxygen atoms leads to hydroxyl radicals and hydrogen atoms. This reaction is least effective near the tropopause and most effective near the stratopause. This loss of H_2 is balanced not only by an upward transport of air through the tropopause but also by a production of H_2 in the mesosphere due to a reaction between HO_2 and atomic hydrogen, H. This latter production of H_2 in the mesosphere has a pronounced maximum near the mesopause. Thus, the major sink of H_2 occurs near the stratopause with sources above (reactions near the mesopause) and below (transport through the tropopause) this altitude. The vertical distribution of H_2 in the stratosphere will, therefore, be highly dependent on vertical transport processes.

Recent measurements (Scholz et al., 1970) indicate a high mixing ratio of H_2 at the stratopause level. This finding will be significant for the related content of H and H_2 in the thermosphere (Bates and Nicolet, 1965). Measurements of both H_2 and H_2O in the stratosphere will provide insight into the destruction of H_2 and H_2O by oxidation in the stratosphere and by photodissociation of H_2O in the mesosphere.

9.5.7
Nitrous Oxide

Nitrous oxide, N_2O, is also subject to photodissociation in the stratosphere (Bates and Hays, 1967). The main source of N_2O in the stratosphere is transport from the troposphere where its concentration is 2.5×10^{-7} (Schütz et al., 1970). A possible *in situ* stratospheric source is formation in a three-body collision between a nitrogen molecule and an excited oxygen atom (Nicolet and Vergison, 1970). The presence of N_2O in the stratosphere leads to the production of nitric oxide (NO) (Nicolet and Vergison, 1970). Excited oxygen atoms react rapidly (rate coefficient of 2×10^{-10} cm^3 sec^{-1}, Greenberg and Heicklen, 1970) with N_2O in either of two ways: first to yield two NO molecules and second, with almost exactly the same frequency, to produce O_2 and N_2 molecules.

9.5.8
Higher Nitrogen Oxides and Nitric Acid

Nitrogen oxide, NO, and nitrogen dioxide, NO_2, are usually lumped together as oxides of nitrogen (NO_x). There are several possible sources of nitrogen oxides in the stratosphere including upward transport from the troposphere (Cadle and Allen, 1970). The second, involving the reaction of N_2O with excited O atoms, was noted earlier. Third, a downward flux of NO can come from the ionosphere above 120 km through the mesosphere without being completely photodissociated (Nicolet, 1970). Three natural sources, therefore, can contribute to the NO_x in the stratosphere. The two oxides are always coexistent in the daytime (Nicolet, 1965). In the upper stratosphere NO is always more abundant than NO_2, but in the lower stratosphere their concentrations are of the same order of magnitude.

NO has not yet been detected in the stratosphere. NO_2 has been measured (Ackerman and Frimout, 1971) by infrared spectroscopy using a spectral line at 3.45 μ, yielding a mixing ratio whose upper limit is 3×10^{-8}. The simultaneous presence of OH, HO_2, NO, and NO_2 leads to a rapid formation of nitric acid,

HNO_3 (Nicolet, 1965), and, in fact, HNO_3 has been detected by infrared absorption spectroscopy in the stratosphere (Murcray et al., 1969). Its mixing ratio according to a laboratory calibration of Murcray's stratospheric solar spectra appears to be of the order of 5×10^{-9} near 20 kilometers (Rhine, Tubbs, and Williams, 1969).

9.6
Particles in the Stratosphere
9.6.1
Nature and Distribution of Stratospheric Particles

Both the scientific and the popular literature are filled with accounts of the vivid sunsets seen all over the world following the major eruption of Krakatoa in the East Indies in 1883. The explanation of the time, set forth in a book published by the Royal Society in 1889, hinged on the effects of a cloud of small dust particles injected into the high atmosphere that spread over the entire earth. Cronin (1971) and Mitchell (1970) have extended this reasoning to the more recent eruption of Agung in 1963, Awu in 1966, Fernandina in 1968, and others, showing that these too must have succeeded in injecting dust, volcanic gases, and entrained air into the lower stratosphere.

In 1960 and 1961 Junge and his collaborators (1961) were among the first to obtain direct samples of stratospheric dust particles by balloon ascents. The particles tended to have a maximum concentration near 18- to 20-km altitude and consisted predominantly of sulfate particles. Other more recent observations by directly sampling (Cadle et al., 1969, 1970), by balloonborne optical measurements (Rosen, 1969), by twilight measurements of scattering (Meinel and Meinel, 1967; Volz, 1969, 1970), and by many laser radar ("lidar") observations have verified Junge's findings and have shown that the concentration of particles in this lower stratospheric layer can vary markedly over a period of years.

After the Agung eruption there was a marked increase in particle concentration—for about two years the sunsets were noticeably more brilliant, reminiscent of the post-Krakatoa displays, but the most recent lidar and direct sampling observations show a gradual decrease toward the pre-Agung values. The Agung dust cloud, plus additions from subsequent eruptions, caused a 30-fold increase in particle mass in 1968 over that in 1961. The amount of filterable material found in 1968 to 1970 by the flights of the National Center for Atmospheric Research in the United States was about 30 to 40 $\times 10^{-4}$ ppm by mass. The measurements of direct sunlight at the high-altitude station on Mauna Loa (see Figure 9.9) have shown a similar decrease in 1963 followed by a gradual climb back toward the earlier values, a change probably attributable to stratospheric particles.

The finding that the stratospheric particles are predominantly sulfate originally came as a surprise. The current explanation is that they are produced from sulfur dioxide gas, SO_2, that is transported upward, mostly in the tropics (or is injected by volcanic explosions), and is subsequently oxidized in the stratosphere to SO_3 by photochemical reactions with O_3 and O. The SO_3 so formed almost immediately hydrolyzes to form sulfuric acid, H_2SO_4, and the samples show that the particles are usually tenths of micron-size droplets of H_2SO_4 containing relatively small amounts of such cations as ammonia and alkali metals (Cadle et al., 1970).

In the course of time, apparently within about two years, these particles grow by coagulation with each other and fall out by sedimentation or are transported out of the stratosphere at middle or high latitudes. The estimate of one- to two-year residence time comes from studies of the rate of depletion of radioactive debris injected into the lower tropical stratosphere by nuclear tests. An injection at mid-latitude would have a shorter residence time, probably less than one year (Martell, 1971).

In a qualitative way this general scheme accounts for the increase of particle density with altitude up to the sulfate layer maximum, since the particles are formed and grow in size as they are transported upward. The remarkable thinness of the layer on occasion (2 or 3 km) is still not adequately explained.

9.6.2
Effects of Stratospheric Particles on Temperature

The influence of the sulfate and other particles in the stratosphere on the solar radiation has already been mentioned, for example, the decrease in the direct solar radiation observed at Mauna Loa

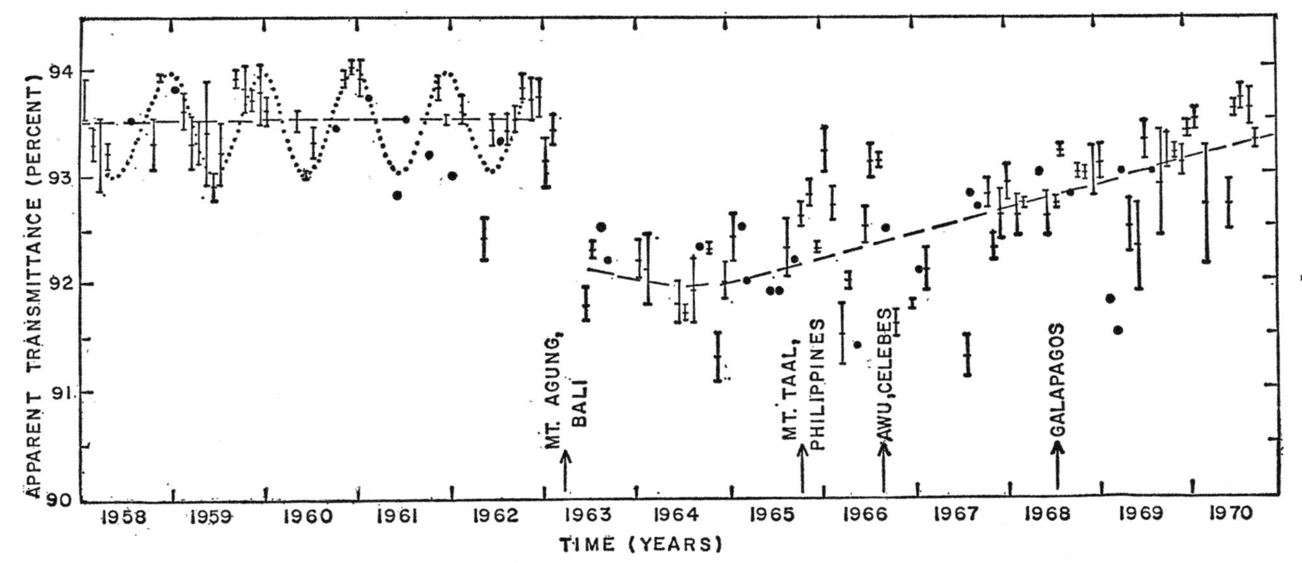

Figure 9.9 Transmittance of normal incidence solar radiation at Mauna Loa, Hawaii (19°N). The dotted line between 1958 and 1962 is a suggested seasonal variation in transmittance. The dashed lines are best-fit curves to the transmittance data displayed as either bar graphs giving range of uncertainty or as heavy dots. Analyses are confined to morning observations. Times of major volcanic eruptions appear along the bottom of the figure.
Source: Ellis and Pueschel, 1971.

following the Agung eruption. Such a decrease would be expected, since particles scatter and absorb solar radiation. The *total* attenuation due to the combined effects of scattering and absorption on the direct solar radiation can be measured. Following the Agung eruption in 1963 it was about 5 percent (Flowers and Viebrock, 1965); for Krakatoa in 1883 it was about 10 percent, and for Katmai in 1912 it was more than 20 percent as measured throughout Europe (Budyko and Pivovarova, 1967; Budyko, 1970). Despite these measurements, it has been very difficult to estimate the absorption that would cause heating of the stratospheric layer. Calculations by Yamamoto and Tanaka (1971) using a range of reasonable absorption and scattering coefficients show that the Agung layer could have absorbed enough solar radiation to cause about 0.08°C per day rise in temperature. Estimates indicate that this could result in a new equilibrium temperature of the order of several degrees higher than that without the volcanic particles. Such a change is consistent with the temperature rises that were observed in 1963 at heights above about 15 kilometers in the tropical stratosphere (SCEP, 1970; Newell, 1970).

There has been some dispute about whether the observed temperature rise in the stratosphere following Agung was due solely to the radiational effects of the particles, since there was a change in the circulation pattern of the stratosphere at the same time. For over a decade prior to 1963 the winds in the tropical stratosphere at heights above about 20 km had been reversing their zonal flow every 26 to 28 months, the phase of this change being such that the reversal progressed slowly downwards. This is known as the "quasi-biennial oscillation." In 1963 the expected change at 20 kilometers did not occur, and it has been suggested (Sparrow, 1971) that the observed warming could have a dynamical explanation.

A decrease in the direct solar radiation following several major recent eruptions has been cited, ranging from 5 to 20 percent. The attenuation of total radiation reaching the ground has been much less, however, and Budyko (1971) estimates that the reduction of total ("global") radiation is less than that of the direct radiation by a factor of 0.13 at the equator, by 0.24 in polar regions, and by 0.16 for a global average corresponding to 30°

latitude. This means that the Agung stratospheric particles probably reduced the total solar radiation at the surface by less than 1 percent. There have been no reports of temperature changes at the earth's surface after the Agung eruption. The cataclysmic eruptions of Krakatoa in 1883 and Mount Tambora in 1815 may, however, have had a significant though transitory effect on global surface temperature. It is quite likely that there have been periods of prolonged volcanic activity in the past that have resulted in appreciable cooling of the lower atmosphere, resulting in a true climate change. (See Section 3.3.6, especially Figure 3.4.)

9.6.3
Possible Man-made Additions of Stratospheric Particles

There are two ways by which man could influence the particle concentration of the stratosphere: by directly injecting them or their gaseous precursors from high-flying aircraft; and by changing the SO_2 content of the troposphere enough to modify the sulfate layer as tropospheric air is transported upwards. We shall treat each of these briefly.

The contributions from high-flying aircraft were analyzed by SCEP and are summarized in Section 9.2 of this chapter.

The second possible man-made stratospheric influence is an accelerating production of SO_2 by the burning of sulfur-rich coal without a corresponding control of sulfur emissions. It is now estimated that man's injection of sulfur in the form of SO_2 into the atmosphere, a large portion of which subsequently becomes sulfate before being removed by natural processes, is one-third to one-half that of nature's production of atmospheric sulfur compounds, and by the year 2000 A.D. man will about match nature if the present trend continues (Robinson and Robbins, 1970; Kellogg et al., 1971). Thus, the nonvolcanic SO_2 that moves upwards into the stratosphere can be increased by a substantial amount, perhaps 50 percent. The fact that a sulfate layer existed in 1960 to 1961, following a period of little volcanic activity, indicates that there must be a background concentration of stratospheric sulfate particles from SO_2 released at the surface. Man can add to this background level. This would still be very much smaller, probably by an order of magnitude, than the concentrations following an eruption such as Agung or Fernandina.

9.6.4
Conclusions

The persistent layer of particles centered at 18 to 20 kilometers in the stratosphere, consisting predominantly of sulfate particles, has shown very marked enhancement following major volcanic eruptions. This enhancement has reduced the direct sunlight reaching the surface by a substantial amount (5 to 20 percent), but has reduced the total solar radiation reaching the surface by less than one-fifth as much (about 1 to 4 percent). Such a reduction must influence the heat balance of the upper troposphere but is probably strongly damped in its propagation downward to the surface. At any rate, the persistence of a cloud from a single eruption is only one to three years, so the effect on surface temperature from eruptions in the past century has been transitory. For past periods of greater volcanic activity, however, the temperature effects may have been more persistent. Volcanic particles in the stratosphere produce temperature rises there due to absorption of sunlight.

A source of man-made stratospheric particles is SO_2 principally from coal combustion that is transported into the stratosphere. During periods of low volcanic activity this man-made addition may increase the background of stratospheric particles by about 50 percent by 2000 A.D., but this would probably be unimportant for the radiation balance of the atmosphere.

9.7
General Conclusions

The sensitivity of the stratosphere to even small intrusions of man-made substances follows from the long residence time within the stratosphere and from the possible role of even small concentrations of certain substances in the ozone photochemistry and heat budget. The stratosphere contains the greater part of the ozone that shields life from potentially damaging ultraviolet radiation; this service for the biosphere is performed by concentrations of the order of parts per million. The ozone also plays a role in the stratospheric heat budget. For these two important reasons it is vital to be able to explain the existing ozone concentration and to predict future changes from man's or nature's intrusion of pollution into the stratosphere. It is not now possible to provide a comprehensive explanation for the observed ozone distribution for three reasons: (1) not all the constituents of the stratosphere are known, (2) some of the chemical reaction rates are still uncertain, and (3) atmospheric transport of ozone and other ozone-producing and ozone-destroying trace gases have received too little attention. No adequate forecast of future ozone changes due to man's activities can be expected until all three impediments are eliminated.

Despite the high cost of sampling minute amounts of trace gases at high altitudes, the requirement for a monitoring program justifies the effort. There is a need for new methods and in some cases new instrumentation. The creation of a suitable model of atmospheric transport is far from trivial; it may be more difficult than sampling the high atmosphere.

The problem of stratospheric particles is primarily one of describing their numbers and characteristics. Since volcanoes inject massive amounts into the stratosphere, a program of monitoring following the next eruption will allow nature to perform an experiment that may help to assess atmospheric response to man's injection of particles into the stratosphere. Further, this, together with routine monitoring during periods of low volcanic activity, will provide the background against which man's contribution will be judged.

Clouds in the stratosphere are rare. Without further information on their occurrence and character it is impossible to assess the role that more or less stratospheric moisture can play in their modification. In particular, they should be sought in regions of the stratosphere with temperatures lower than about −75°C.

9.8
Recommendations

1. We recommend that the upper troposphere (above about 10 km) and lower stratosphere (up to at least 30 km) be monitored for the following constituents: (1) water vapor, oxides of nitrogen, including nitrous oxide, oxides of sulfur, hydrocarbons, including methane, (2) particles, determining insofar as possible the sizes, concentrations, chemical composition, and optical charac-

teristics, and (3) stratospheric cloudiness. Special consideration should be given to the following:

a. Develop a method for sensing water vapor in the parts per million range based on a principle other than frost point hygrometry. The sensor must also be absolute.

b. The sensors for oxides of nitrogen and sulfur and for the hydrocarbons (other than methane) should be developed with the capability of detecting concentrations in the parts per billion range.

c. Particle detectors capable of sensing the presence of particles as small as 0.01 to 0.1 μ in radius should be developed.

d. The spectroscopic method that can detect ozone, water vapor, carbon monoxide and carbon dioxide, nitrous oxide, nitrogen dioxide, nitric acid, and methane as well as other unidentified constituents be employed aboard platforms to at least 30 kilometers in order to obtain simultaneous concentrations and vertical distributions of all constituents.

2. We recommend that special efforts to monitor the stratosphere be undertaken when and if another major volcanic eruption injects debris into the stratosphere.

3. We recommend that the absolute values of ultraviolet solar flux involved in ozone-oxygen reactions be determined with an accuracy of at least 5 percent to ensure proper calculations of photo-oxidation rates in the stratosphere. New and more precise measurements are needed both at stratospheric levels and above the atmosphere.

4. We recommend the analysis in laboratory simulations of those rate coefficients needed for calculations of the stratosphere photochemistry which are not known to a factor of 2. We also recommend that the absorption cross sections of such minor constituents as hydrogen peroxide, H_2O_2, hydroperoxyl, HO_2, nitric acid, HNO_3, nitrous acid, HNO_2, and nitrogen dioxide, NO_2, be measured.

5. We recommend that a research program be instituted at the earliest possible date to analyze the photochemical-dynamical interaction for stratospheric ozone, with special emphasis on exhaust products from high-flying aircraft.

6. We recommend that theoretical modeling and tracer studies

designed to understand the average transfer and changes in the rate of transport of tropospheric air into the stratosphere be undertaken.

9.9
Some Implications for Implementation of the Recommendations

Recommendation 1 calls for an extended series of measurements in the upper troposphere and stratosphere which would involve the use of high-altitude aircraft, balloons, and aircraft-mounted remote-sensing devices. It calls for accurate measurements of concentrations of some materials that are near the lower limit of sensitivity of current detecting technique. For several constituents new techniques of detection and measurement are required. It is a difficult and costly program. The full material implications are hard to assess, but the cost will be high. Nevertheless, it will be a very small fraction of the sum already spent on the development of supersonic aircraft.

Recommendation 2 calls for airborne and surface observations organized in such a way as to allow immediate intensification when a major volcanic eruption throwing debris into the stratosphere occurs.

Recommendation 3 calls for an increase in current capabilities in the very difficult field of absolute measurement of ultraviolet radiance. Sophisticated aircraft and balloon facilities are required, but probably not too many flying hours are needed. These facilities exist and would probably be made available. Measurements on artificial (and perhaps the natural) satellites are also necessary. The sensors, when developed, are likely to be comparable with existing vehicles and payloads.

Recommendations 4 and 5 and 6 call for theoretical, computational, and laboratory work, all of high caliber and some very difficult, but all are feasible with current techniques.

References

Additional references for Chapter 9 are on page 295.

Ackerman, M., and Frimout, D., 1969. Mesure de l'absorption stratosphérique du rayonnement solaire de 3.05 à 3.70 microns, *Bulletin de l'Académie Royale de Belgique, Cl. Sc., 55:* 948–953.

Auerbach, H., 1961. Geographic variation in incidence of skin cancer in the United States, *Public Health Reports, 76:* 345–348.

Bainbridge, A. E., and Heidt, L. E., 1966. Measurement of methane in the atmosphere and lower stratosphere, *Tellus, 18:* 221.

Bates, D. R., and Hays, P. B., 1967. Atmospheric nitrous oxide, *Planetary and Space Science, 15:* 189.

Bates, D. R., and Nicolet, M., 1950a. Photochemistry of atmospheric water vapor, *Journal of Geophysical Research, 55:* 301–327.

Bates, D. R., and Nicolet, M., 1950b. Atmospheric hydrogen, *Publications of the Astronomical Society of the Pacific, 62:* 106–109.

Bates, D. R., and Nicolet, M., 1965. Atmospheric hydrogen, *Planetary and Space Science, 13:* 905.

Bischof, W., and Bolin, B., 1966. Space and time variations of CO_2 content of the troposphere and lower stratosphere, *Tellus, 18:* 155–159.

Blum, H. F., 1959. *Carcinogenesis by Ultraviolet Light* (Princeton, New Jersey: Princeton University Press).

Brewer, A. W., 1949. Evidence for a world circulation provided by measurements of helium and water vapor distributions in the stratosphere, *Quarterly Journal of the Royal Meteorological Society, 75:* 351–363.

Budyko, M. I., and Pivovarova, Z. I., 1967. Influence of volcanic eruptions on the incoming surface solar radiation, *Meteorologia i Hydrologia, 11:* 301.

Cadle, R. D., 1964. Daytime atmospheric $O(^1D)$, *Discussions of the Faraday Society, 37:* 66.

Cadle, R. D., and Allen, E. R., 1970. Atmospheric photochemistry, *Science, 167:* 243.

Cadle, R. D., Lazrus, A. L., Pollock, W. H., and Shedlovsky, J. P., 1970. The chemical composition of aerosol particles in the tropical stratosphere, *Proceedings of the American Meteorological Society Symposium on Tropical Meteorology,* edited by C. Ramage (Honolulu, Hawaii: Institute of Geophysics).

Chapman, S., 1930. A theory of upper atmospheric ozone, *Memoirs of the Royal Meteorological Society, 3:* 103–125.

Chapman, S., 1943. Photochemistry of atmospheric oxygen, *Reports of Progress in Physics, 9:* 92.

Craig, R., 1965. *"The Upper Atmosphere" Meteorology and Physics* (New York and London: Academic Press).

Cronin, J. F., 1971. Recent volcanism and the stratosphere, *Science, 172:* 847–849.

Crutzen, P. J., 1969. Determination of parameters appearing in the dry and wet photochemical theories for ozone in the stratosphere, *Tellus, 21:* 368–388.

Crutzen, P. J., 1970. The influence of nitrogen oxides on the atmospheric ozone content, *Quarterly Journal of the Royal Meteorological Society, 96:* 320–325.

Crutzen, P. J., 1971. Ozone production rates in an oxygen-hydrogen-nitrogen oxide atmosphere (unpublished).

DeMore, W. B., 1967. New mechanism for OH-catalyzed chain decomposition of ozone, *Journal of Chemical Physics, 46:* 813.

Dütsch, H. U., 1968. The photochemistry of stratospheric ozone, *Quarterly Journal of the Royal Meteorological Society, 94:* 483–497.

Ellis, H. T., and Pueschel, R. F., 1971. Solar radiation: absence of air pollution trends at Mauna Loa, *Science, 172:* 845–846.

Flowers, E. C., and Viebrock, H. J., 1965. Solar radiation: An anomalous decrease of direct solar radiation, *Science, 148:* 493–494.

Gebhart, R., Bojkov, R., and London, J., 1970. Stratospheric ozone: a comparison of observed and computed models, *Beiträge zur Physik der Atmosphäre, 43:* 209–227.

Greenberg, R. I., and Heicklen, J., 1970. Reaction of $O(^1D)$ with N_2O, *International Journal of Chemical Kinetics, 2:* 185.

Greiner, N. R., 1970. Hydroxyl radical kinetics by kinetic spectroscopy, *Journal of Chemical Physics, 53:* 1070.

Hampson, J., 1966. Chemiluminescent emissions observed in the stratosphere and mesosphere, *Les Problèmes Météorologiques de la Stratosphère et de la Mésophère,* edited by Marcel Nicolet, pp. 393–440.

Hays, P. B., and Olivero, J. J., 1970. Carbon dioxide and monoxide above the troposphere, *Planetary and Space Science, 18:* 1729.

Hesstvedt, E., 1958. Mother-of-pearl clouds in Norway, *Geofysiske Publikasjoner, 10:* No. 10.

Hesstvedt, E., 1968. On the photochemistry of ozone in the ozone layer, *Geofysiske Publikasjoner, 27:* 1–16.

Hesstvedt, E., 1970. Vertical distribution of CO near the tropopause, *Nature, 225:* 50.

Hunt, B. G., 1966. Photochemistry of ozone in a moist atmosphere, *Journal of Geophysical Research, 71:* 1385–1398.

Hunt, B. G., 1969. Experiments with a stratospheric general circulation model-III large-scale diffusion of ozone including photochemistry, *Monthly Weather Review, 97:* 287–306.

Johnston, H., 1971. Reduction of stratospheric ozone by nitrogen oxide catalysts from SST exhaust, *Science, 173:* 517–522.

Junge, C. E., and Manson, J. E., 1961. Stratospheric aerosol studies, *Journal of Geophysical Research, 66:* 2163–2182.

Junge, C., Seiler, W., and Warneck, P., 1971. The atmospheric ^{12}CO and ^{14}CO budget, *Journal of Geophysical Research, 76:* 2866.

Kaufman, F., 1964. Aeronomic reactions including hydrogen, *Annales de Géophysique, 20:* 106.

Kellogg, W. W., Cadle, R. D., Allen, E. R., Lazrus, A. L., and Martell, E. A., 1971. Man's contributions are compared to natural sources of sulfur compounds in the atmosphere and oceans (unpublished).

Komhyr, W. D., Barrett, E. C., Slocum, G., and Weickmann, H. K., 1971. Atmospheric total ozone increases during the 1960s, *Nature* (in press).

Langley, K. F., and McGrath, W. D., 1971. The ultra-violet photolysis of ozone in the presence of water vapor, *Planetary and Space Science, 19:* 413–416.

Leovy, C. B., 1969. Atmospheric ozone: An analytic model for photochemistry in the presence of water vapor, *Journal of Geophysical Research, 74:* 417–426.

London, J., 1968. The average distribution and time variation of ozone in the stratosphere and mesosphere, *Space Research, 7:* 172–184.

London, J., 1969. The depletion of ultraviolet radiation by atmospheric ozone, *Biologic Effects of Ultraviolet Radiation,* edited by Frederick Urbach (Oxford: Pergamon Press Ltd.) pp. 335–339.

McKinnon, D., and Morewood, H. W., 1970. Water vapor distribution in the lower stratosphere over North and South America, *Journal of the Atmospheric Sciences, 27:* 483.

Martell, E. A., 1971. Residence times and other factors influencing pollution of the upper atmosphere, *Man's Impact on the Climate,* edited by W. H. Matthews, W. W. Kellogg, and G. D. Robinson (Cambridge, Massachusetts: The M.I.T. Press), pp. 421–428.

Mastenbrook, J. H., 1971. The variability of water vapor in the stratosphere (unpublished).

Meinel, A. B., and Meinel, M. P., 1967. Volcanic sunset glow stratum: Origin, *Science, 155:* 189.

Mitchell, J. M., 1970. A preliminary evaluation of the atmospheric pollution as a cause of the global temperature fluctuation of the past century, *Global Effects of Environmental Pollution,* edited by S. F. Singer (Dordrecht, Holland: Reidel Publishing Company, and New York: Springer-Verlag), p. 139.

Murcray, D. R., Kyle, T. G., Murcray, F. H., and Williams, W. J., 1969. Presence of HNO_3 in the upper atmosphere, *Journal of the Optical Society of America, 59:* 1131.

Newell, R. E., 1970. Modification of stratosphere properties by trace constituent changes, *Nature, 227:* 697–699.

Nicolet, M., 1965. Nitrogen oxides in the chemosphere, *Journal of Geophysical Research, 70:* 679–689.

Nicolet, M., 1970a. Aeronomic reactions of hydrogen and ozone, *Aeronomica Acta,* A-No. 79 (Brussels: Institut d'Aéronomie Spatiale de Belgique).

Nicolet, M., 1970b. The origin of nitric oxide in the terrestrial atmosphere, *Planetary and Space Science, 18:* 1111–1118.

Nicolet, M., 1970c. Ozone and hydrogen reactions, *Annales de Géophysique, 26:* 531–546.

Nicolet, M., and Vergison, L., 1970. Nitrous oxide in the stratosphere, *Aeronomica Acta,* A-No. 89 (Brussels: Institut d'Aéronomie Spatiale de Belgique).

Prabhakara, C. P., 1963. Effects of non-photochemical processes on the meridional distribution and total amount of ozone in the atmosphere, *Monthly Weather Review, 91:* 411–431.

Pressman, J. A., and Warneck, P., 1970. The stratosphere as chemical sink for carbon monoxide, *Journal of the Atmospheric Sciences, 27:* 155.

Preston, K. F., and Cvetanovic, R. J., 1966. Collisional deactivation of excited oxygen atoms in the photolysis of NO_2 at 2288 Å, *Journal of Chemical Physics, 45:* 2888.

Rhine, P. E., Tubbs, L. D., and Williams, D., 1969. Nitric acid vapor above 19 km in the earth's atmosphere, *Applied Optics, 8:* 1500.

Robinson, E., and Robbins, R. C., 1970. Gaseous atmospheric pollutants from urban and natural sources, *Global Effects of Environmental Pollution,* edited by S. F. Singer (Dordrecht, Holland: Reidel Publishing Company, and New York: Springer-Verlag), pp. 50–64.

Rosen, J. M., 1969. Stratospheric dust and its relationship to meteoric influx, *Space Science Reviews, 9:* 59–89.

Study of Critical Environmental Problems (SCEP), 1970. *Man's Impact on the Global Environment* (Cambridge, Massachussetts: the M.I.T. Press).

Scholz, T. G., Ehhalt, D. H., Heidt, L. E., and Martell, E. A., 1970. Water vapor, molecular hydrogen, methane, and tritium concentration near the stratopause, *Journal of Geophysical Research, 75:* 3049.

Schütz, K., Junge, C., Beck, R., and Albrecht, B., 1970. Studies of atmospheric N_2O, *Journal of Geophysical Research, 75:* 2230.

Seiler, W., and Junge, C., 1969. Decrease of carbon monoxide mixing ratio above the polar tropopause, *Tellus, 21:* 447–449.

Sissenwine, N., Grantham, D. D., and Samela, H. A., 1968. Mid-latitude humidity to 32 km, *Journal of the Atmospheric Sciences, 25:* 1129.

Trosko, J. E., Krause, D., and Isoon, M., 1970. Sunlight-induced pyrimidine dimers in human cells *in vitro, Nature, 228:* 358–359.

Urbach, F., 1969. Geographic pathology of skin cancer, *The Biologic Effects of Ultraviolet Radiation* (Oxford: Pergamon Press), pp. 635–661.

Volz, F. E., 1969. Twilights and stratospheric dust before and after Agung eruption, *Applied Optics, 8:* 2505–2517.

Volz, F. E., 1970. Atmospheric turbidity after the Agung eruption of 1963 and size distribution of the volcanic aerosol, *Journal of Geophysical Research, 75:* 5185–5193.

Williamson, E. J., and Houghton, J., 1965. Radiometric measurements of emission from stratospheric water vapour, *Quarterly Journal of the Royal Meteorological Society, 91:* 330.

Yamamoto, G., and Tanaka, M., 1971. Increase of global albedo due to air pollution (unpublished).

Units and Conversion Factors

LENGTH

0.001 kilometer (km) = 1 meter (m) = 100 centimeters (cm) = 1,000 millimeters (mm)
1 meter (m) = 1,000,000 microns (μ) = 1,000,000,000 nanometers (nm) = 10,000,000,000 angstrom units (Å)
1 km = 0.6214 statute mile; 1 m = 39.37 inches (in.) = 3.281 feet (ft); 1 cm = 0.3937 in.
1 mile = 1.609 km; 1 ft = 0.3048 m; 1 in. = 2.54 cm

WEIGHT

0.001 kilogram (kg) = 1 gram (g) = 1000 milligrams (mg) = 1,000,000 micrograms (μg)
1 kg = 2.205 pounds (lb); 1 g = 0.035 ounce (oz)
1 lb = 453.6 g = 0.4536 kg
1 metric ton = 2205 lb = 1000 kg
1 short ton = 2000 lb = 907.2 kg
1 megaton = 1,000,000 tons

VOLUME AND CUBIC MEASURE

1 cubic meter (m^3) = 1,000,000 cubic centimeters (cm^3)
1 m^3 = 35.31 cubic feet (ft^3); 1 cm^3 = 0.061 cubic inch ($in.^3$)
1 liter (l) = 1000 cm^3
1 l = 61.02 $in.^3$ = 0.2642 gallon (gal)
1 ft^3 = 0.02832 m^3 = 28.32 l; 1 gal = 231 $in.^3$ = 3.785 l
1 unit = 1,000,000 parts per million (ppm) = 1,000,000,000 parts per billion (ppb)

SOLAR RADIATION

1 Langley/day = 0.485 watt/m^2 (W/m^2)

ENERGY CONSUMPTION DENSITY

1 kilowatt-hour/day (kWh/day) = 0.86 Mcal/day

ENERGY AND WORK

1 British thermal unit (Btu) = 252 calories (cal) = 0.0002931 kilowatt-hour (kWh)
1 Wh = 860.421 cal = 3.413 Btu

POWER

1 megawatt (MW) = 1000 kilowatts (kW) = 1,000,000 watts (W) = 3,413,000 Btu/h = 1341 horsepower (hp)

PRESSURE

1 atmosphere (atm) = 76 cm mercury = 14.70 lb/$in.^3$ = 1013 millibars (mb)

TEMPERATURE SCALES

	Absolute Zero	Ice Point (water)	Steam Point (water)
Degrees Fahrenheit (°F)	−459.7	32	212
Degrees Celsius or Centigrade (°C)	−273.15	0	100
Degrees Kelvin (°K)	0	273.15	373.15
Degrees Reaumur (°R)		0	80

TIME

1 Ma = 10^6 years

Additional References

Chapter 3

Dansgaard, W., Johnsen, S. J., Møller, C., and Langway, C., Jr., 1969. One thousand centuries of climatic record from Camp Century on the Greenland ice sheet, *Science, 166:* 377–381.

Lamb, H. H., 1968. Volcanic dust, melting of ice caps, and sea levels. *Palaeogeography, Palaeoclimatology, Palaeoecology, 4:* 219–222.

Mitchell, J. M., Jr., 1970. A preliminary evaluation of atmospheric pollution as a cause of the global temperature fluctuation of the past century. *Global Effects of Environmental Pollution,* edited by S. F. Singer (Dordrecht, Holland: Reidel Publishing Company; New York: Springer-Verlag), pp. 139–155.

Pütz, F. R., 1971. Klimatologische Wetterkarten der Nordhemisphäre für die Temperatur in Dezennium 1961/70, *Beilage zur Berliner Wetterkarte, 106:* 23.

Rasool, S. I., and DeBergh, C., 1970. Runaway greenhouse and the accumulation of CO_2 in the Venus atmosphere, *Nature, 226:* 1037–1039.

von Rudloff, H., 1967. *Schwankungen und Pendelungen des Klimas in Europa seit Beginn regelmässiger Beobachtungen* (Braunschweig: Vieweg).

Chapter 8

Abel, N., Jaenicke, R., Junge, C., Kanter, H., Rodriguez Garcia Prieto, R., and Seiler, W., 1969. Luftchemische Studien am Observatorium Izana (Teneriffa), *Meteorologische Rundschau, 22:* 158–167.

Arrhenius, S., 1896. The influence of the carbonic acid in the air upon the temperature of the ground, *Philosophical Magazine, 41:* 237–276.

Atwater, M. A., 1971. The radiation budget for polluted layers of urban environment, *Journal of Applied Meteorology, 10:* 205–214.

Beckwith, W. B., 1971. Private communication to W. W. Kellogg.

Bigg, E. K., 1965. Problems in the distribution of ice nuclei, *Proceedings of the International Conference on Cloud Physics,* Tokyo and Sapporo, pp. 137–140.

Flohn, H., 1971. Comments on Peterson and Junge estimates on sources of atmospheric particles (unpublished).

Junge, C., 1960. Sulfur in the atmosphere, *Journal of Geophysical Research, 65:* 227–237.

Junge, C., and Ryan, T. G., 1958. Study of the SO_2 oxidation in solution and its role in atmospheric chemistry, *Quarterly Journal of the Royal Meteorological Society, 84:* 46–55.

Keeling, C. D., 1971, in press.

Kelley, J. J., Jr., 1969. An analysis of carbon dioxide in the arctic atmosphere near Barrow, Alaska, 1961–1967 (University of Washington), Report #NR 307–252.

Koenig, L. R., 1968. Some observations suggesting ice crystal multiplication in the atmosphere, *Journal of Atmospheric Sciences, 25:* 460–463.

Squires, P., 1966. An estimate of the anthropogenic production of cloud nuclei, *Journal des Recherches Atmosphériques, 2:* 297–308.

Twomey, S., 1971, personal communication.

Went, F. W., 1966. On the nature of Aitken condensation nuclei, *Tellus, 18:* 549–556.

Yamamoto, G., Tanaka, M., and Asano, S., 1970: Radiative transfer in water clouds in the infrared region, *Journal of Atmospheric Sciences, 27,* 282–292.

Chapter 9

Budyko, M. I., 1971. *Climate and Life* (Leningrad: Hydrological Publishing House).

Cadle, R. D., Bleck, R., Shedlovsky, J. P., Blifford, I. A., Rosinski, J., and Lazrus, A. L., 1969. Trace constituents in the vicinity of jet streams, *Journal of Applied Meteorology, 8:* 348–356.

Georgii, H., and Jost, D., 1969. Concentration of CO_2 in the upper troposphere and lower stratosphere, *Nature, 221:* 1040.

Jost, D., 1971. Untersuchung der CO_2-Konzentration im Tropopausenniveau, *Pure and Applied Geophysics* (PAGEOPH), *86:* 209–218.

Index